Genetics of Flowering Plants

Genetics of Flowering Plants

R.C. Upadhyaya

ANMOL PUBLICATIONS PVT. LTD.
NEW DELHI-110 002 (INDIA)

ANMOL PUBLICATIONS PVT. LTD.
H.O.: 4374/4B, Ansari Road, Darya Ganj,
New Delhi-110 002 (India)
Ph.: 23278000, 23261597
B.O.: No. 1015, Ist Main Road, BSK IIIrd Stage
IIIrd Phase, IIIrd Block
Bangalore - 560 085 (India)
Visit us at: www.anmolpublications.com

Genetics of Flowering Plants

First Published, 2008

ISBN 978-81-261-3428-1

PRINTED IN INDIA

Printed at Mehra Offset Press, Delhi.

Contents

Preface

Horticulture has a special significance for the economic development. Most of the people, though hard working and face the calamities of nature and live in harsh conditions, yet produce lot of fruits, vegetables and other flowering and ornamental plants including the medicinal and aromatic plants. In order to increase the productivity of horticultural crops growers must take the advantage of the expertise of scientists. Keeping this idea in focus, the present work has been compiled which contains experiences of the experts and related people. The volumes cover advances in strategies, production, plant protection, value addition and other important areas of horticulture development.

The Encyclopaedia provides practical information on production technologies of tropical and subtropical fruits as well as plantation crops and horticultural nursery. Post-harvest management, marketing and export trends have been furnished for guidance. It also presents most realistic cost estimates and returns from cultivation of such crops. The volumes aim at viewing commercial prospects for horticulture. It has been carefully designed to meet the queries of intending investors, financial institutes, creditors, traders and growers engaged in dealing with horticultural crops. Further it will serve the purpose reference books to students, academicians and related people.

R.C. Upadhyaya

1

Flowers and Foliage Plants

Flower and foliage plants are some of the most colourful and attractive horticultural plants. Some are used to decorate the rooms in our homes while others are used to beautify the areas around our homes and public buildings. Because winter temperatures seldom reach freezing in south Florida, cold sensitive foliage plants normally grown as house plants in north and central Florida can be grown as landscape plants in south Florida.

The purpose of this section of the study manual is to promote a better understanding of flowers and foliage plants. It is primarily designed to aid in the identification of the more common flowers and foliage plants grown in Florida. Plants are arranged alphabetically by common names. Common names are used for the convenience of persons not accustomed to using scientific names. Scientific names are in parentheses immediately following common names. Scientific names are the same throughout the world with plant scientists and seldom change.

The scientific name is usually composed of two words. The first word is always capitalized and is called the genus. The second word is usually not capitalized and is called the species. There may be many species of plants in one genus which differ only slightly in appearance. When reference is made to several species in one genus, the abbreviation "spp." is often written after the genus name.

When a group of plants within a species differ in one or more characteristics from their parent plants, they are classified into another category. When this occurs naturally, the new category is called a variety and we add it to the scientific name, such as Dianthus plumarius var. albiflorus. When new varieties appear while plants are in cultivation, they are called "cultivars" (cultivated varieties). The cultivar name is also a part of the scientific name and may be listed such as Dianthus chinensis cv. Splendens or Dianthus chinensis `Splendens'. Single quote marks will be used in this publication to designate cultivars.

The plants in this section of the study manual are described using many botanical terms. These terms are used to describe types of leaves, leaf position and arrangement, leaf shapes, leaf margins, flower parts and flower arrangement. It may be helpful, when studying these plant descriptions, to refer to the glossary of terms located in the rules and glossary section of the Florida 4-H Horticulture Identification and Judging Study Manual. Space is provided for you to write notes that help you identify the plant. You may also want to cut out photographs from old nursery magazines and catalogues to glue or tape in this manual to further help you in identification.

African Violet (Saintpaulia spp.): A small plant with a rosette form. Leaves and stems are very pubescent. Leaves can be oval to heart-shaped with smooth or ruffled edges. Flowers are borne in clusters, can have 5 or more petals and come in different shades and combinations of white, blue and red.

Aglaonema (Aglaonema commutatum): An herbaceous perennial with attractive foliage variously marked depending on the cultivar. Thick stems are sheathed by leaf petioles which are shorter than the blades. The flowers are small spadices with small spathes which form showy, bright red or yellow fruits in clusters.

Aloe Vera (Aloe barbadensis): A succulent plant with stiff rosettes composed of leaves which are usually mottled with purple or white; they have spiked edges or may be smooth. The flowers are showy tubular blossoms borne on spikes in tones of yellow. The sap is used for soothing skin burns.

Amaryllis (Hippeastrum spp.): A low growing, stemless herb with tunicate bulb. Simple leaves are basal and entire, linear to strap-shaped to 2 long. They are glabrous, light green and arise with or after flower stalk emerges. Anywhere from 1-10 umbellate flowers may be produced on a hollow leafless stalk to 2 tall. Flowers are large and funnel-shaped to 10 across. The prevailing colours are red, pink, white and often lined or striped. Flowering occurs in the winter and spring.

Begonia (Begonia spp.): An erect\or trailing, sometimes climbing herbaceous annual or perennial. Usually succulent in texture. Leaves are simple, alternate, highly variable in size and shape, characteristically asymmetrical, glabrous or pubescent. Male and female flowers are usually borne in axillary cymes, some quite showy. They come in a variety of sizes, shapes and colours.

Caladium (Caladium x hortulanum): A tuberous, stemless perennial herb with colourful leaves, which are simple, basal and entire, ovate to lanceolate shaped; may be up to 14 long. Petioles are longer than the leaf blades. Caladiums show a variety of variegated colours: red, rose, white, green, and pink. The flower is an inconspicuous spadix and spathe.

Canna (Canna x generalis): A tall, perennial, erect herb with thick branching rhizomes. Simple leaves are alternate to spirally arranged, with entire margins and sheathing petioles. They are ovate to elliptic-lanceolate in shape, often with prominent veins. Irregularly shaped flowers are borne in spikes, and may or may not be tubular at the base. The fruit is a rough, three-valved capsule.

Celosia (Celosia spp.): A showy garden plant grown as an annual. Flowers are tiny and crowded into a feather-shaped spike or flattened crest (cockscomb). Flower colour varies from red to yellow. Leaves are alternate.

Christmas Cactus (Schlumbergera brigesii): No foliage is present in this perennial cactus. The stems are flat and segmented. The margins of each segment are rounded and scalloped. Flowers are borne at ends of stems are very showy, in shades or red, white or orange.

Chrysanthemum (Chrysanthemum x morifolium): A much-branched, erect, herbaceous perennial with simple leaves which are alternately arranged. The leaves are lobed, entire to coarsely toothed, lanceolate to ovate, often pubescent and thick, with a strong fragrance. Flowers are heads of various sizes and shapes, clustered with disk and ray florets. Colours are many shades of white, yellow, pink, bronze, and red purple.

Coleus (Coleus x hybridus): A freely-branching, densely-foliated herbaceous perennial. Simple leaves are opposite with margins which may be crenate, serrate, undulate, lobed, etc. They are generally ovate and soft textured, They exhibit many colours and colour combinations; patterns of pink, white, yellow, red, green, and maroon are common. The stem is squared in cross section, green or brown, becoming woody with age.

Daylily (Hemerocallis spp.): A clumping perennial herb, spreading by fleshy rootstocks which may be evergreen or deciduous. Leaves are more or less linear, wider at the base and tapering gradually to the tip. The flowers are 6-petaled, funnel form to campanulate, and are born in clusters on long scapes. They come in shades of yellows, orange, red and purple, and bloom in the spring and/or summer.

Dianthus (Dianthus chinensis): There are many species of dianthus. This species, commonly called China pink, grows 6 to 12 inches in height and produces an enormous quantity of 1- to 2- inch single, semidouble or frilled flowers in shades of pink, rose, scarlet, crimson and white. The foliage is gray-green in colour.

Dieffenbachia (Dieffenbachia spp.): A herbaceous evergreen perennial with thick stems. The large leaves are alternate, usually with white markings or patterns which have a very striking appearance. The petioles are long and cover the stem. The flowers are in spadices with thick spathes.

Dracaena Marginata (Dracaena marginata): Evergreen, multiple-stemmed plant with fine textured appearance. It has sword or strap-shaped leaves which are crowded into tufts at the ends of branches. These leathery leaves have a red stripe along the edges of the entire margins.

Episcia (Episcia spp.): A herbaceous, creeping perennial with very pubescent leaves in dark tones of copper, mostly variegated with silver and green. The small flowers are showy red, purplish or white.

Geranium (Pelargonium x hortorum): Leaves are orbicular, undulate and coarsely crenate with a spicy scent. Flowers are in rounded umbels four to six inches across. The individual flowers are five-petaled. They can be red, pink or white.

Gladiolus (Gladiolus x hortulanus): This perennial herb grows from a corm and has erect, unbranched flower stalks. Simple leaves are linear, sword-shaped or cylindrical. They are often rigid with a somewhat papery texture and have prominent parallel veins. Flowers are borne on one sided spikes; they are irregular with 6 segments united into a curved funnelform tube. The upper three segments are larger that the lower three. Flowers range in colour from yellow, white, orange, red, lavender-purple, to brown. The fruit is a three-valved, large, flattened or winged capsule.

Gloxinia (Sinningia speciosa): These are tuberous plants with oblong-ovate dentate leaves, 6 to 8 inches long. Flowers are solitary or few clusters. They are bell-shaped to 3 inches long, usually violet or purple, but may be reddish or white spotted.

Grape Ivy (Cissus rhombifolia): This evergreen vine has shiny, sharply toothed, three-lobed leaves, 4 inches long and 3 to 4 inches wide, that unfold bronzy, then turn deep green. New buds and the undersides of mature leaves are covered with soft brown fuzz. Flowers are rare and inconspicuous.

Heart-leaf Philodendron (Philodendron scandens subsp. oxycardium): An evergreen herbaceous vine with simple entire leaves climbing by aerial roots. Leaves are glossy-green, cordate (hence the common name "heart-leaf philodendron") and are arranged alternately. The petioles are round and do not clasp the stem. The dry leaf sheath remains around the stem.

Holly Fern (Cyrtomium falcatum): This fern has a dense spiral crown with an arching growth habit. Fronds may be up to 2½ feet long, are compound, and are arranged spirally.

Leaflets are ovate to elliptic-shaped to 5 long, and may be entire, weakly wavy or toothed; they are very glossy and waxy on the surface and lighter coloured beneath, with a leathery texture. Spores on the undersides are very distinct and look like scale insects.

Impatiens (Impatiens spp.): An erect, spreading, succulent annual which is very brittle. Simple leaves are glabrous, alternate, sometimes opposite with serrate margins. They are green or reddish-green, ovate-lanceolate to elliptic-oblong in shape. Flowers are solitary and borne on axillary or terminal racemes and are in many forms, doubled, semi-doubled, etc. Flowers have a spur which contains nectar and they come in many colours: purple, variegated whites, yellows, reds, and oranges.

Jade Plant (Crassula argentea): A small shrub with thick, succulent trunk and branches, resembling a small tree. Leaves are opposite, obovate and lacking petioles, attached directly to the stem. They are very thick, succulent and often have red edges.

Marigold (Tagetes spp.): A densely-foliated, freely-branching, erect, herbaceous annual, with odd-pinnately compound leaves, which are opposite with serrate margins, lanceolate. They have glandular dots on the undersides, and are strongly aromatic. Flowers are dense heads to 5 across, solitary, borne on long peduncles. They have few to many ray florets, often two-lipped or quilled, in yellows, oranges, reddish-browns or multi-coloured.

Neanthe Bella Palm (Chamaedorea elegans): A small, fine-textured, single trunked palm, usually 4 high. Leaves are pinnately compound; leaflets are lanceolate, with short, unarmed petioles. Fruits are black and globose, maturing throughout the year.

Nephthytis (Syngonium podophyllum): A creeping herbaceous perennial attaching itself to surfaces by means of aerial roots. Leaves have long petioles, are spear shaped and marbled with white while young, becoming very large and deeply lobed when old.

Norfolk Island Pine (Araucaria heterophylla): An evergreen coniferous tree with a single, upright trunk; the branching habit is tiered and canopy is pyramidal in shape. Leaves are dark green, soft and spirally arranged. The trunk is rough and old branches fall off, leaving distinct scars.

Pansy (Viola x wittrockiana): This winter flowering perennial is grown as an annual in Florida. Pansies grow to a height of 8 inches. The leaves are dark green, drooping, with wavy edges. They have delicately fragrant 2- to 3-inch flowers of five overlapping petals looking like gigantic violets, except that the colours are purple, white, blue, dark red, rose or yellow combined in almost endless variations of stripes and blotches.

Peperomia (Peperomia spp.): An herbaceous perennial with a clumping growth habit. The succulent leaves are quite variable in size, shape and colour. Leaves are frequently shiny and ovate. Flower spikes are small and inconspicuous.

Petunia (Petunia x hybrida): A densely-foliated, trailing, low-growing herbaceous annual. Simple leaves are alternate with upper leaves sometimes opposite; margins are entire, and shape varies from linear-oblong to spatulate; texture is soft and pubescent. Flowers are axillary and solitary to 5 across, with funnelform corollas which are often deeply fringed or full double. They come in many colours and combinations of whites, yellows, reds and blues.

Poinsettia (Euphorbia pulcherrima): Large, upright growing shrub which can reach a height of 12 feet. Leaves are large, alternate, ovate and light green in colour. The showy portion of the plant, popularly referred to as the flower, consists of modified leaves, which are called bracts. Bracts may be red, pink, yellow or white. One cultivar has red bracts with pale pink spots and blotches. The true flowers are small, green and yellow nubs clustered in the centres of the bracts.

Pothos (Epipremnum aureum): A vigorous, evergreen herbaceous vine with simple, alternate leaves which are ovate to cordate. Juvenile leaves are entire, but mature leaves are perforated and can reach a lenght of 30 inches. The petiole is channeled at the base and clasping the stem. They are glossy, bright green or irregularly variegated with lighter green shades or white. The juvenile form is the most commonly seen.

Prayer Plant (Maranta leuconeura): A low-growing perennial plant with colourful leaves, noted for the beautiful patterns of lines, blotches and spots on leaves. They vary in colour from soft, light green to dark green on the upper surface and light green on the undersides. They are called "Prayer Plants" because of their habit of folding their leaves at night.

Salvia (Salvia spp.): Herbaceous annual, usually with square stems. Simple leaves are opposite, oval or lanceolate. Margins are toothed or segmented. Flowers are borne in clusters, growing from the axils of small bracts which vary in colour from scarlet, purple, blue to white and pale yellow.

Sansevieria (Sansevieria spp.): A stiff, erect, herbaceous perennial with short and thick rhizomes. Leaves are elongate, very thick, erect, flat or cylindrical and are usually borne basally. They are often mottled or variegated. Flowers though rarely produced and then only on older plants, are arranged in racemes and are often fragrant.

Schefflera (Brassaia actinophylla): An upright evergreen tree, often with multiple trunks. It is low-branching with palmately compound spirally arranged leaves of 7-15 leaflets which are oblong. The leaflets radiate from the apex of an elongated petiole. Leaves are entire or sparsely dentate on the juvenile plant and are glossy green.

Snapdragon (Antirrhinum majus): An erect herbaceous annual or perennial with stiff stems. The leaves are simple and alternate, with entire margins. They are lanceolate to oblong-lanceolate. The stems are usually covered with short, sticky hairs. Flowers are borne on terminal recemes, growing from the axils of small leafy bracts. The corolla is tubular to 2 long and pouched, forming a mouth. Colours are red, pink, yellow, orange, and purple with all shades and combinations of these.

Spathiphyllum (Spathiphyllum spp.): Clumping herbaceous perennial with large dark-green, plantain-like leaves which are thin, lanceolate in shape. Leaf petioles are as long or longer than the blade. Flowers are long spadices with white or greenish spathes. The fruits are fleshy and berry-like.

Spider Plant (Chlorophytum comosum): An herbaceous perennial with linear leaves which are bright, smooth, and often lined with white bands.

They are arranged in rosettes, and bunches of small plantlets are produced at the end of long stolons. The flowers are small, white, to ¾ across, borne in racemes, and precede the plantlets.

Swedish Ivy (Plectranthus australis): A low-growing herbaceous perennial with broadly ovate, simple, fleshy leaves which are arranged oppositely on square stems. The margins are pronouncedly crenate. Flowers are arranged in racemes, and are small and pale purple.

Verbena (Verbena x hybrida): Evergreen herbaceous perennial with a creeping growth habit. Leaves are simple, opposite, ovate to ovate-oblong and margins are toothed.

Its fragrant flowers are borne in flattened umbels on quadrangular stems. Flower colours are pink, red, white, yellow, blue and purple.

Weeping Fig (Ficus benjamina): This evergreen tree has drooping, spreading branches which are heavily foliated. The leaves,which exude a milky sap, are simple and entire.

They are alternate, and ovate-elliptic in shape and leathery-textured with a glossy-green colour. The stem is grey, and brown aerial roots eventually appear.

Zebra Plant (Aphelandra squarosa): Shiny, dark green opposite leaves on one to two foot stems with conspicuous white stripping along the veins. White, 2-lipped flowers emerge from yellow bracts in clusters atop the main stem.

Zinnia (Zinnia elegans): An erect, stiff-stemmed herbaceous annual. Simple leaves are oppositely arranged with entire margins, usually sessile or clasping the stem. They are lanceolate, ovate, or oblong in shape, dark green, pubescent with prominent veins.

Flowers are solitary heads, borne terminally. Disc and ray florets are often elongated, twisted and tubular. They come in every colour but blue, some are multi-coloured.

The Lawn Grasses

There are four ways to plant a new lawn: seed, sod, sprigs, plugs. Seed is the traditional way, and has the advantage of relatively low cost. If you're planting a large area seed is probably the way to go. Over the last three decades of so, turfgrass sod has become widely available. Somewhere, somehow, most grasses are available as sod. Sod is a good choice when the planting area is smaller, or where time or resources to nurse a seedbed until germination.

Think of sprigs and plugs as pieces of sod: they're essentially the same as sod, but divided into smaller units. Plugs are actually small squares of sod, and sprigs are pieces of mature lawn grass that can form roots and grow. There are a few lawn grasses, such as hybrid Bermuda, St. Augustine, and some zoysia varieties, that are available only as sod, plugs, or sprigs.

Here are brief descriptions of the different kinds of lawn grasses. The kinds that you can plant as seed include bahia, bent, Bermuda, blue grama, buffalo, centipede, fescue (all kinds), bluegrass, ryegrass (both annual and perennial), and some zoysias.

Bahia grass (Paspalum notatum) is a tough, coarse grass that is popular in the Southeast as a low maintenance and generally low quality lawn. It is drought and shade tolerant, but requires frequent mowing. Improved varieties such as 'Pensacola' are more cold-tolerant and somewhat more attractive. Seashore paspalum (P. vaginatum) has a fine-textured leaf blade, but is most notable for its tolerance of salty soil and water.

Bent grass, creeping (Agrostis stolonifera) is a fine-textured perennial grass that forms a soft, tightly-knit turf. A favourite for golf course putting greens in northern regions. Top varieties include 'Penneagle', 'Penncross', and 'Seaside'.

Bermuda grass (Cynodon dactylon) is the most important lawn grass of the Sun Belt. It has all the essential characteristics a lawn needs for those regions-heat and drought tolerance primarily. The very aggressiveness of Bermuda is also it's key disadvantage. When creeping Bermuda invades flower beds and ground cover, it's called "devil grass" by frustrated gardeners. Bermuda grass is tough and fast growing enough to be one of the most popular grasses for sports fields. It's good for pet owners for the same reasons: damage from whatever cause will repair quickly. Bermuda tolerates little shade and will thin quickly in shade; and it becomes dormant and brown through cool seasons.

There are many varieties of Bermuda grass. Common Bermuda or "Arizona common" is widely available. It's a relatively coarse-bladed form. Medium textured varieties available as seed include 'Cheyenne', 'Guymon', 'Jackpot', 'NuMex Sahara', 'Sonesta', 'Sundevil', 'U3', and 'Yuma'. The softest and finest blade Bermuda-the kinds used for southern and tropical golf greens-are hybrids that are available as sod only. These include 'Tifgreen', 'Tiflawn', 'Tifway'.

Blue grama grass (Bouteloua gracilis) is a North American native grass that is well suited to the northern plains. Like Bermuda grass, it grows best in summer and is drought tolerant. But unlike Bermuda grass, it is very cold hardy and able to survive a North Dakota winter. Blue grama spreads slightly, but ultimately produces a clumpy, uneven lawn. Mowing it 3 or 4 times a year will make it more lawn like.

Buffalograss (Buchloe dactyloides) is, like blue grama, a North American native grass, a once dominate grass of the vast American short-grass prairie. For obvious reasons, it makes good sense as a lawn grass in much of the low-rainfall West. The look is similar to Bermuda: slightly blue-green in the summer, then straw brown through winter. Several improved varieties are available; some sod only. The best seed varieties are 'Cody' and 'Tatanka'. A close runner-up is 'Topgun'. Top varieties to grow from sod are '609', 'Prairie', and 'Stampede'.

Centipede grass's (Eremochloa ophiuroides) prime virtue is its ability to survive and even thrive in acidic, poor soils. Because temperatures below 5°F kill centipede grass, only gardeners in the southeastern United States (and parts of Hawaii) grow it. A warm season grass, it turns brown in hot and dry weather, and is quick to enter dormancy in fall.

Fescue, fine- or needle-leaved. There are three types: chewings fescue (Festuca rubra commutata), creeping red fescue (F. rubra), and hard fescue (F. longifolia). Chewings fescue is desirable in low maintenance situations. It is particularly tolerant of sandy, acidic, and infertile soils. Hard fescue is very cold hardy and low maintenance. A naturally short grass, it requires less frequent mowing. Creeping red fescue is used most widely. All three are often mixed with Kentucky bluegrass and perennial ryegrass, adding greater shade and drought tolerance. Named varieties include 'Aurora', 'Bighorn', 'Claudia', 'Flyer', 'Ensylva', 'Longfellow,' 'Marker', 'Medallion', 'Reliant', 'Salem', 'Scaldis', 'Shademaster', 'Spartan', and 'Victory'.

Fescue, tall or broad-leaved (Festuca arundinacea). Traditional varieties, such as 'Kentucky 31' and 'Alta', are tough and coarse-bladed pasture grass. But now many improved varieties are available that nearly match Kentucky bluegrass in look and feel. Tall fescues are more heat and drought tolerant compared to fine fescues. Varieties to look for include 'Amigo', 'Avanti', 'Aztec', 'Bonsai', 'Cimarron', 'Cochise', 'Crewcut', 'Crossfire', 'Duster', 'Guardian', 'Monarch', 'Ninja', 'Pixie', Rebel Jr.', 'Shenandoah,' 'Tomahawk', 'Tribute', 'Twilight', 'Virtue', and 'Wrangler'. Some-such as 'Rebel III', 'Earth Saye', 'Shenandoah', 'Titan II', and 'Tarheel'-are endowed with natural pest repellents, called endophytes.

Kentucky bluegrass (Poa pratensis). By far the most popular type of lawn grass for most of the northern half of the country. It is also planted further south, usually in areas with significant coastal influence, such as coastal southern California. It grows there, but long term survival is rare. Notably, it does not thrive in the Pacific Northwest, west of the Cascade Mountains. Cold-hardy, soft-textured, attractive dark green colour. Look for named varieties-such as 'A-34', 'America', 'Asset', 'Blacksburg', 'Blue Star', 'Chateau', 'Classic', 'Eclipse', 'Glade', Julia', 'Loft's 1757', 'Midnight', and 'Princeton 104'-on seed label.

Ryegrass, annual (Lolium multiflorum). Annual ryegrass has one redeeming virtue: its low cost. For that reason it is often used to overseed dormant warm-seasons grasses in southern regions. Because it prefers cool temperatures, most will die out of a warm-season lawn once hot summer days arrive.

Ryegrass, perennial (Lolium perenne). The best varieties of perennial ryegrass have a desirable fine, soft texture, and dark green colour of Kentucky bluegrass. While occasionally used alone, they are most often combined with Kentucky bluegrass and fine fescues. Look for varieties such as 'Commander', 'Dimension', 'Manhattan II', 'Palmer', 'Pennant', 'Pick 715', 'Riviera', 'Saturn', 'SR-4000', and 'SR-4100'. 'Manhattan II' is notable for being endowed with natural pest repellents, called endophytes.

Seashore paspalum (Paspalum vaginatum) is a native of South African sand dunes and is well adapted to similar environments in the United States It is noted for its tolerance of salty soil and irrigation water. It can survive on ocean water, and is well suited to partially recycled gray water. Its look and feel is similar to Bermuda grass, and it similarly spreads with both above- and below-ground runners. It is planted by sprigs, plugs, or sod.

St. Augustine grass (Stenotaphrum secundatum) is one of the most important grasses of the South and West. It's fast-growing, deep-rooted, and coarse-textured with broad, blunt-tipped blades. It is a spreading grass via above-ground runners. Seed of this grass is not available; plant it from sprigs, plugs, or sod only. Top varieties are 'DelMar', 'Floralawn', and 'Jade'.

Zoysia grass (Zoysia species) is often touted as a miracle grass, and while many of the claims of its virtues are true, there are also drawbacks. Three species available: Japanese lawn grass (Z. japonica), Manila grass (Z. matrella), and Korean grass (Z. tenuifolia). Korean grass is more a ground cover than lawn grass.

It's wiry but fine-textured and creates interesting mounds as it grows. Only Japanese lawngrass is available as seed. While tough and resilient once established, it is notorious for its poor germination, short green season, and long dormant season. The newest varieties 'Zen 300' and 'Zenith', improve upon all three traits. Hulled and treated seed germinate in

2 weeks and can make a lawn in 2 months. The colour of Japanese lawngrass is similar to bluegrass, but blades are much stiffer. Vegetative selections of Japanese lawngrass include 'Meyer' and 'El Toro'. Manila grass has stiff and flat leaves with a fine texture and deep green colour. It makes a high quality lawn in tropical areas, but is planted from sprigs, plugs, or sod only. Other zoysias to plant in the same way include 'Cashmere', 'De Anza', 'Emerald', and 'Victoria'.

An Introduction to Organic Gardening

Today's average homeowner has become accustomed to using dozens of potentially harmful chemicals to treat plant pests and diseases in the yard and garden. Many gardeners also rely on chemical fertilizers and pesticides to grow their grass, flowers, fruits, and vegetables.

Unlike large scale farms, which tend to measure and use fertilizers and pesticides as they are intended, many homeowners overuse pesticides and fertilizers, hoping that if a little bit is good, then more must be better.

According to the National Academy of Sciences, even though farmers uses pesticides more widely, homeowners uses 10 times more per acre.

Chemicals we use in our yards can, over the long run, negatively affect the environment and the health of our loved ones, neighbours, and pets. Particularly if they are used incorrectly.

Also, many of the chemical fertilizers and pesticides can degrade the health of our soil, garden plants, lawns, and trees over time, making the use of strong chemicals even more necessary just to keep our plants alive.

Once you kill off all the beneficial organisms, then you are stuck with sterile, dead dirt and chemicals instead of rich, organic soil.

With that in mind, we hope to introduce you to some organic gardening techniques and products that apply to the average homeowner who isn't necessarily a gardening or yard care professional or fanatic.

Organic gardening can serve as a healthy alternative to the use of chemicals in the yard and garden.

Perhaps you don't think that it's entirely necessary or practical for you to stop using all chemical pesticides, fertilizers, and herbicides completely in your yard. By adding a few environmentally friendly techniques and products in your daily gardening routine, you can save money by reducing the amounts of chemicals and pesticides that you need.

If you plan on continuing with chemicals, please follow the instructions and don't overuse them! It doesn't help to use more than recommended, and it can definitely hurt you and your lawn.

Even if you adopt just a few of these garden practices, you can have big impacts on your yard, the environment and the health of your family and neighbours.

And by using these organic gardening techniques, you will over time create a stronger, healthier garden that will require less maintenance.

After a few gardening seasons, you might find that organic gardening techniques are easier and more effective than you ever imagined.

What's Wrong with Chemical Gardening Products?

While most chemical gardening products pose limited risk if used correctly, many of these products actually kill helpful insects and soil organisms and can damage a plant's natural ability to fend off pests and diseases.

Chemical fertilizers can also build up in the soil and eventually reduce productivity. As the soil quality gradually declines, more fertilizer is necessary to replace vital nutrients that should be naturally present in your garden soil.

Additionally, chemicals used in the lawn and garden have been scrutinized by a variety of academic and scientific studies. The results don't exactly inspire confidence in the pesticides, herbicides and fertilizers many of us routinely use in our yards. These products have been linked to birth defects, cancer, and environmental degradation.

Organic Gardening as an Alternative to Chemicals

Organic gardening is a method of gardening that has existed for centuries. It wasn't until the advent of chemical gardening products that organic gardening began to be pushed aside as an "alternative" or "luxury" form of gardening. As this happened, those who practiced organic gardening had to redefine themselves. Gardeners generally define organic gardening as working to create a natural balance of healthy soil and healthy plants in the garden. We can look to nature for ideas on how to grow healthy plants and how to utilize more natural processes instead of chemicals to keep our gardens healthy.

Organic gardening focuses on building up the soil, using native plants and plants appropriate for your garden, and looking for a natural balance in your garden. Pathogens generally attack weak plants that are not properly adapted to their environment, and those that live in poor soil.

So, if we work our soil and encourage the presence of beneficial soil organisms, our plants will generally be stronger and more resistant to pests and diseases. When we integrate organic techniques into our gardening, we'll help to ensure cleaner neighbourhoods, cleaner water, and healthier families.

Recommended Gardening Products and Techniques

Organic gardening methods are constantly evolving. New technologies and new techniques are coming out every day. We've selected some of the most basic and effective organic gardening techniques in this article.

Remember that one of the great things about organic gardening techniques is that they apply to any kind of garden you grow, from flower gardens, to herb gardens, to your vegetable garden.

Organic Gardening Concepts and Techniques

Composting : Composting is one of the best ways to build up the soil, attract beneficial soil organisms, improve soil drainage, and provide your plants the nutrients that they need to thrive. Please check out the Compost Guide and our Composting section for information and composting products.

Mulching: Mulch reduces evaporation rates, adds nutrients to the soil, keeps weeds at bay, and protects your soil from erosion. A 3 to 4 inch layer of mulch is ideal for your garden beds. Replace the mulch as needed.

Grasscycling : Leaving grass clippings on your lawn will actually improve the health of your turfgrass.

Native Plants : Native plants are naturally adapted to your area, which means that they need less water and less maintenance. They are also more resistant to native pests. Try and use at least a few native plants in your garden. Learn more about native plants here.

Biodiversity: A diverse garden is a happy garden. Using a wide variety of plants in your garden means a higher number of beneficial insects and soil organisms. Biodiversity also means that pests are less likely to eat up all your plants! Read more about biodiversity here.

Companion Planting : Companion planting is the concept of using certain plants in combinations to keep pests away or otherwise strengthen plants.

Recommended Organic Gardening Products

Mycor Root Builders: This product stimulates the growth of beneficial Mycorrhizal fungi in the soil. It works with vegetables, ornamentals, grass, etc.

Organic Insecticides: Natural citrus-based insecticides are popular with many gardeners and are effective at killing off most of the pests you're likely to see in your garden. It's safe enough to use indoors. Beer is also a natural way to kill off slugs and snails, which are attracted to the scent and then drown in a glass or specially designed slug and snail trap.

Hot Pepper Wax: Hot Pepper Wax sprays help to keep both animals and insects away from edible crops. A food grade wax helps the oils stay on your plants, without leaving a spicy residue.

Bat Guano: Bat guano has a high humus content and works great as a soil builder and fertilizer. It's also 100 percent natural, from the butts of bats. Look for bat guano that meets the standards of Bat Conservation International.

Organic Fire Ant Killer: Red imported fire ants can be a serious problem in the yard and garden. Try out a natural fire ant killer to put an end to these pesky critters. You can use a fire ant killer as step one of the Texas A&M developed Texas Two Step fire ant control method

Citrus and Vinegar-Based Organic Weed Killers: Natural citrus and vinegar-based liquids are very effective weed killers and cause the weeds to wilt and die within minutes. You may need a couple treatments for tough perennial weeds.

Organic Liquid Lawn Fertilizer: For a nice, lush yard without chemicals, try out an organic liquid lawn fertilizer. Look for products that are a combination of lawn food supplements, natural soil activators, and Humate Liquid Humus.

Horticultural Corn Meal: Horticultural cornmeal is an organic way of controlling harmful soil fungi and problems with plant roots. This is especially important for vegetable crops.

Corn Gluten Meal : Corn gluten meal is a natural herbicide that also fertilizes the soil. Please visit Dr. Nick Christians' University of Iowa Corn Gluten research page for more information on the benefits of corn gluten meal.

Still looking for more good information? Keep reading, with our Buyer's Guide for Organic Pest Control.

Safe Lawn and Garden Campaign

The Safe Lawn and Garden Campaign is a joint project of EAGLE and the Duluth Community Garden Programme funded by the Minnesota Pollution Control Agency. The goal of this campaign is to educate businesses and homeowners about options for simple, nontoxic methods of lawn and garden care.

Lawn Care

Many homeowners look for quick fix solutions to lawn and garden "problems" on the shelves of their local hardware store, garden centre, or other retailer. They assume if the product is for sale to the public it must be safe to use. However, out of 30 commonly used pesticides, 13 are probable or possible carcinogens, 13 are linked with birth defects, 21 with reproductive effects, 15 with neurotoxicity, 26 with liver or kidney damage, 27 are sensitizers and/or irritants and 11 have the potential to disrupt the endocrine (hormonal) system.

Pesticides

Pesticides are also a danger to pets, fish, birds, beneficial insects, and other wildlife. Chemicals do not necessarily stay on the ground where they are applied. Thirteen lawn pesticides have been found in ground-water or surface water in the Great Lakes Basin.

Fertilizers

Synthetic, quick release fertilizers are also a problem as they run off into area streams and lakes, causing excess algae growth that robs other aquatic life of oxygen. Additionally, overusing high nitrogen, water-soluble fertilizers can contribute to dangerous levels of nitrates forming in drinking water.

This has been linked to Methemoglobinemia, also known as Blue Baby Syndrome. Methemoglobinemia is a condition in which nitrates turn to nitrites (NO_2) in babies' digestive tracts, and reduce the ability of hemoglobin to carry oxygen. It can cause death in infants.

On top of the many possible environmental and health hazards, chemically treating your lawn or garden also weakens the soil, and creates a cycle of dependency that can be costly and time consuming.

Safe Lawns

By only using nontoxic methods of lawn and garden care you are protecting yourself, your family, your neighbours, and the entire ecosystem. Join others across the nation who are pledging to go pesticide free! Please take a moment to sign our Safe Lawn and Garden Pledge.

For a limited time, Duluth area residents who sign our pledge will receive at no cost, Pesticide Free Lawn sign. Signs can be picked up at EAGLE's office in the DeWitt Seitz Building in Canal Park or alternative pick up arrangements can be made.

Less is More: A Lawn Care Primer

Many Americans accept the idea that we must have lawns of thick, green, healthy grass with few weeds. We spend a lot of time, money, and energy trying to achieve that perfectlawn. All too often, we direct our efforts toward trying to discourage the plants and insects we do not want by using chemical weed killers called herbicides.

Herbicides are a type of pesticide. Other types include insecticides and fungicides. Pesticides cause significant health and environmental problems. They negatively affect people, pets, and wildlife.

The easier and healthier way to maintain a green lawn is to encourage the plants we do want. By optimizing the conditions that encourage grass growth, it is possible to have a healthy, green lawn without using chemicals. The Safe Lawn and Garden Campaign promotes alternative lawn management practices that eliminate the use of pesticides and synthetic fertilizers.

Less Mowing

Don't mow your lawn too short. Maintain a height of 3 to 3 1/2". Mowing your grass at the proper height is the single most important thing you can do to improve the health of your lawn.

Mow when your lawn needs it. Don't mow every weekend just because it's tradition, and don't let the grass grow so long that you have a hay field. Maintain a fairly constant height by mowing when the grass is about 4" tall, so that you're always removing about 1/3 of the leaf blade. In Duluth this usually means more frequent mowing in spring and fall, less frequent mowing in summer.

Always mow with sharp mower blades so the grass is cut cleanly, not ripped.

Why are these Things Important?

The depth of the grass root is proportional to the height of the grass blade. By letting the grass grow, you allow the roots to grow down. Deeper roots can reach more water during dry periods, making it less necessary to water the grass.

Grass manufactures food from sunlight. The longer the blade, the more food the plant can make, keeping it healthier and stronger. The deeper roots can also reach more nutrients.

Longer leaf blades create more shade, making it harder for weeds to get established.

Less Watering

Here in the Northland we usually get enough rain during the summer so that it is not necessary to water the lawn. In addition, grasses that survive best in the Great Lakes Region naturally slow down their growth during the hot, dry part of the summer. During this dormant period they lookbrown and dead, but they are only resting. Maintaining your grass at least 3" high will help it to survive dry periods without extra watering.

If you do water your lawn:

- Water only if it has not rained for 7 days.
- Water early in the morning—between 6 and 10 a.m.
- Thoroughly soak the ground, giving it about 1" of water.

Less Raking

If you leave the grass clippings on the ground as fertilizer, you will not need to rake your lawn except to clean up fallen leaves. Grass clippings decompose rapidly and do not promote disease or thatch buildup. If you rake the lawn in spring, be sure to rake gently so you do not pull up on the young grass plants.

No Pesticides

Using pesticides to control weeds is like treating symptoms of a cold. You may succeed in killing the weeds, but you will not change the conditions that allowed the weeds to grow. A weed-free lawn is not necessarily a healthy lawn. Weeds can tellyou something about what is wrong with your lawn.

Here in Minnesota:

- Plantain may indicate the soil is compacted or poorly drained.
- Creeping Charlie may indicate the site is too shady or the soil is poorly-drained.

- Hawkweed may indicate the soil is low in nutrients.
- Dandelions may indicate the grass is too thin.
- Moss may indicate the site is too shady or too wet for grass to survive.

The easiest way to control weeds is to optimize the conditions that favour grass. That way the weeds do not have a chance to get established. If you do have weeds, interrupt their cycle of seed production by digging up the roots or cutting off flowering stalks before they go to seed.

Spread a "northern mix" of grass seed in the empty space. After spreading grass seed, dampen it and cover it with mulch. Keep the grass seed evenly damp until it germinates. In heavy shade, plant shade-tolerant ground covers instead of grass.

Less Fertilizing

The soil provides nutrients such as nitrogen, potassium, and phosphorous that are essential for a healthy lawn. Not all soil is equal, but about half of the nitrogen can be provided by leaving grass clippings on the lawn to decompose. Grass clippings break down quickly and do not promote disease or thatch buildup.

If you think your lawn is lacking nutrients, have the soil tested. Call the local County Extension Office (In Saint Louis County, the number is 218-733-2870) to help you determine what kind of fertilizer will correct the deficiencies indicated by the test.

Use slow-release fertilizer, use only the amount needed to correct the deficiencies, and add it in the fall. After you have improved the health of your lawn, you only need to have the soil tested every 3 years.

Keep in mind that "organic" in this sense means made of naturally occurring plant, animal, or mineral material, not synthetic or petroleum based.

Some fertilizers contain toxic chemicals and heavy metals, even ones labelled "organic." When in doubt, use and Organic Materials Review Institute certified product.

Advanced Lawn Maintenance or Giving Your Lawn a Facelift

Many Americans accept the idea that we must have lawns of thick, green, healthy grass with few weeds. We spend a lot of time, money, and energy trying to achieve that perfect lawn. All too often, we direct our efforts toward trying to discourage plants and insects we do not want by using chemical weed killers called herbicides.

Herbicides are a type of pesticide; other types include insecticides and fungicides. Pesticides are poisonous. They can cause sickness in adults, children, and pets, and they can end up in our drinking water.

The easier and healthier way to maintain a green lawn is to encourage the plants we do want. By optimizing conditions that encourage grass growth, it is possible to have a healthy, green lawn without using harmful chemicals. The Safe Lawn and Garden Campaign promotes alternative lawn management practices that eliminate the use of pesticides and synthetic fertilizers.

Basic lawn care practices that improve the health of your lawn without using pesticides include:

- *Mow High :* Maintain a grass height of 3". Allowing the grass to grow tall encourages the roots to go deep, resulting in healthier grass that can withstand dry periods and other stress. This is the most important thing you can do to improve the health of your lawn.
- *Water Sparingly :* Here in the Duluth area we usually get enough rain during the summer so that it is not necessary to water lawns. In addition, grasses that survive the best in the Northland naturally slow down their growth during the hot, dry part of the summer. During this dormant period they look brown and dead, but they are only resting.
- *Don't Rake :* Fertilize by leaving grass clippings on the lawn. If you use a fertilizer, use a slow-release organic variety, use only what is needed, and apply it in the fall.

Remember that grass can not grow successfully everywhere. Just like every other plant, it needs certain physical conditions to survive. The most important conditions for successful grass growth include:

- *Good Soil* : Grass grows best in 4"-6" of well-drained topsoil that includes plenty of organic matter. Soil pH should be between 6.5 and 7.0.
- *Plenty of Sun* : Most varieties of grass need an average of six hours of full sun everyday. If your yard receives less than 6 hours of full sun each day, be sure to plant shade-tolerant grass varieties. If your yard is very shady, consider planting other types of ground cover.
- *Adequate Moisture* : Maintaining your grass at least 3" high will help it to survive dry periods without extra watering.

If you do water your lawn,

- Water only if it has not rained for 7 days.
- Water early in the morning. (Between 6 and 10a.m.)
- Thoroughly soak the ground, giving it about 1" of water.

Improve the Soil

Grass grows best in 4-6" of well-drained soil with plenty of organic material. If your soil is heavy clay or loose sand, add ½" of organic matter in the form of compost. Use a mechanical spreader to distribute the compost evenly over the lawn, then water thoroughly. In addition to adding organic material, this will give earthworms and microorganisms a chance to loosen the soil. A pesticide free lawn is a worm friendly lawn.

Have the pH of your soil tested. The pH is ameasure of the acidity of the soil. Grass does best in soil with a pH of 6.5 to 7.0, which is slightly acidic. If the pH of your soil is lower than 6.5, add agricultural lime or ground limestone to raise the pH. If pH is above 7.0, add sulfur to lower the pH. Contact your local County Extension Office (In Saint Louis County, the number is 218-733-2870) for help in determining how much lime or sulfur to add.

Choose the Right Varieties

To have a healthy lawn, you need to plant grass varieties that will succeed in the conditions found in your lawn. There are many varieties of grasses available today to help you do this. For Northern Minnesota's cool climate and short summers, a mixture of Kentucky bluegrass, perennial rye, and red fescue or tallfescue does the best. You can buy these premixed, usually labelled "Northern mixture." Read the label carefully, and look for a good quality mixture that has no more than 1% weed seed and no more than 1% "unknown" seed.

If your existing lawn is more than 25 years old, it probably consists of Kentucky bluegrass varieties and some fescues. If your lawn is sodded, it is probably composed of improved Kentucky bluegrasses, in a mixture of high maintenance and lower maintenance varieties. Consider over-seeding to introduce lower maintenance varieties and cultivars.

Encourage New Growth

Over-seeding (Sowing grass seeds on top of an existing lawn) can be helpful for revitalizing your lawn. It is a good way to fill in thin spots and a good way to introduce grass varieties that are appropriate for your area. Use a "northern mixture" of cool season varieties of Kentucky bluegrass, perennial rye grass, and red fescue. In the fall use a spreader to distribute the grass seed mixture over the entire lawn. Use about 4-6 pounds of seed per 1000 square feet. Be sure to keep the grass seed evenly moist until it germinates and seedlings are about 1" tall. If the seed dries out, germination success will be substantially lowered. You may need to over-seed more than once before you see a significant change in composition.

Fertilize to Replace Missing Nutrients

Many people fertilize their lawn more often than is necessary. The purpose of fertilizing is to compensate for nutrient deficiency. If your soil is not deficient in nutrients, there is no reason to add fertilizer. To determine whether your lawn does need fertilizer, have your soil tested. Nutrients essential for a healthy lawn include nitrogen (N), phosphorus (P), and potassium (K).

For lower maintenance lawns, grass clippings can provide most of the nutrients your lawn needs. Nitrogen, which is the nutrient required in the largest amounts by grass, is abundant in grass clippings. Grass clippings are 90% water, and they decompose in a week or two. Just by mowing regularly and leaving the clippings on your lawn, you are adding nitrogen to the soil continuously.

Nitrogen (N) encourages growth of green parts. Lack of N causes slow growth, yellow leaves, and thin spots. Too much N causes excessive leaf growth, reduced root growth, low food reserves, and increased susceptibility to environmental stresses. The best source of N for your lawn is grass clippings.

Phosphorus (P) is essential for seedling growth and root development. Lack of P causes reduced growth, dark-green to reddish leaf colour, narrow leaf blades, and wilting. Organic sources of P include manure, compost, grass clippings, and tree leaves.

Potassium (K) is necessary for the development of string stems and disease resistance. Lack of K causes slow growth, weak stalks, and increases susceptibility to environmental stress and disease. Wood ashes contain K, but they also raise the soil pH, souse sparingly.

If you do use fertilizer, do not apply it in the spring. Grass plants put most of their energy into producing flowering stems in the spring. Adding fertilizer that is mostly nitrogen simply stimulates even more stem growth. It does little to encourage root growth, which is essential for the plants to survive during the hot, dry parts of the summer. Fertilizing in the spring also stimulates the growth of weeds.

If you fertilize, choose a slow release variety, and apply it in the fall using a spreader. Applying fertilizer in the fall increases the likelihood that it will be used for root growth, because that is the plant's main fall activity. Having good, strong roots is the best way for the grass plant to stimulate and support vigorous leaf growth in the spring and to survive summer dry periods.

Aerate to Reduce Soil Compaction

Soil compaction is common problem with heavy clay soils. Compacted soil prevents air, water, and nutrients from getting down to the roots of the grass. Indicators of compacted soil include thin grass and plentiful weeds. Aerating your lawn will reduce soil compaction problems. By increasing water and nutrient penetration, aeration stimulates healthy root development. Use a soil aerator, a machine that cuts plugs out of the soil and drops them on the lawn to decompose.

Aerating just once will not make a significant difference to the soil. To be most effective, you will need to aerate 2-3 times at 2-3 week intervals. Start with the first aeration during the third week in July, so that the last one is done in September. After the first aeration, topdress by spreading a thin layer of compost over the entire lawn. Then over-seed by spreading a "northern mixture" of grass seed using about 4-6 pounds of seed per 1000 square feet. Keep the grass seed evenly moist until it sprouts and is about 1" tall. Aerators can be rented at most equipment rental stores, and some lawn care services will do it for you.

Soil Quality

In general, an N-P-K ratio of 3-1-2 is appropriate for a healthy lawn. If the soil test indicates that some of these nutrients are lacking, fertilizer should be added to correct the specific deficiencies. Call your local County Extension Office (In Saint Louis county the number is 218-733-2870) for information on how to perform and interpret your soil test. The regular cost of a soil test is about $15. Once you have improved the health of your lawn, a soil test is probably only necessary every three years.

Some Suggested Lawn Fertilizers

Corn Gluten Meal: Corn gluten meal is a natural, non-toxic "weed and feed" that can reduce the mount of crabgrass and dandelions in your yard while simultaneously fertilizing it. It is pre-emergent, meaning it needs to be applied before weeds go to seed, usually in mid-April, and can be applied again in the fall. It is available in feed stores and garden centres. Please note that corn gluten meal will also prevent the germination of grass seed. If you also wish to over-seed your lawn, do so at least six weeks after applying corn gluten meal.

Remember! If you use a fertilizer, use an organic slow release fertilizer. Keep in mind that "organic" in this sense means made of naturally occurring plant, animal, or mineral materials, not synthetic or petroleum based. Some fertilizers contain toxic chemicals and heavy metals, even ones labelled "organic". When in doubt, use an Organic Materials Review certified product.

Be Flexible

Keep in mind that there is a point of no return. If your existing lawn has more than 50% weeds, it is probably better to till it up and start over. Be sure to choose grass varieties that will thrive in the conditions in your yard. If grass will not do well in some areas of your lawn, consider planting other types of ground cover. Some shade-tolerant ground covers include hosta lilies, ferns, periwinkle, pachysandra, lungwort, ajuga, and barren strawberry.

The Safe Lawn Philosophy: A lawn is a community made up of thousands of individual plants. Each plant in the community competes for the resources it needs—sunlight, water, and nutrients. The easiest way to maintain a healthy lawn is to optimize the conditions that favour grass growth. That way grass can out-compete the weeds, and you will have a green lawn without using pesticides.

Weeds and Other Pests

What is a weed? That depends on your point of view. A weed to one person might be a beautiful wild-flower or a source of food to someone else. You must decide which weeds and how many you are willing to tolerate in your lawn. Remember, a healthy lawn is not necessarily a weed-free lawn.

Most of the plants that we label as weeds are plants that can tolerate poor growing conditions. They thrive in conditions that other plants do poorly in, such as low nutrients or compacted soil. Because they survive in poor growing conditions, weeds can help you diagnose your lawn's problems.

Many weeds are colonizers. They like bare soil and lots of sun, so they are the first to take advantage of newly disturbed ground. Weeds are able to grow quickly and produce huge quantities of seed, some of which can remain viable for decades. Many weeds are plants that have been introduced from other countries. This has given them an advantage over our native plants, because the introduced species have escaped their natural controls like insect, pests, and diseases.

The best way to control weeds is to avoid creating the conditions that encourage weed growth. Follow the Safe Lawn guidelines of mowing high (3") and leaving grass clippings in the lawn. Have your soil tested for pH and nutrients, then add slow release fertilizer in the fall if the soil indicates it is needed. Whenever you remove weeds, remember to spread some grass seed in the bare spot left behind, using a "northern mix" that is appropriate for our climate. Add some annual rye seed to the mix for quick sprouting. Then keep the seeds evenly moist until they sprout.

Permanent weed management lies in nurturing desirable plants the form a multi-layered groundcover within which weeds cannot get established.

Plantain

Characteristics: A perennial with a basal rosette of thick, smooth dark-green leaves. Many small, inconspicuous flowers cover a spike about 6-8" tall.

Indicator of: Compacted or poorly-drained soil.

Control: Dig up and dispose of the plant. Then re-seed the area. Aerate your lawn to loosen compacted soil.

Creeping Charlie

Characteristics: A perennial that grows in sun orshade. A low plant with long creeping square stem and small lavender flowers. Smells like mint when crushed.

Indicator of: Poorly-drained soil and shady conditions.

Control: In the spring, raise the runners by raking, then mow them off. Be sure to dispose of the runners completely, because any pieces left on the ground will simply send out new roots. Or mix 5 teaspoons of borax with 1 quart water. (This will cover a 5' x 5' area.) Spray the mixture directly on the plant when it is in full bloom or after the first frost. Do not apply more than once per season, because it will kill the grass as well.

Hawkweed or Indian Paintbrush (Hieraceum species)

Characteristics: A perennial with a yellow or orange dandelion-like flowers at the top of a stalk about 8-12" tall. The rosettes of hairy leaves often form extensive mats. They spread by underground stems and runners. Hawkweed blooms in summer.

Indicator of: Disturbed ground of low fertility, acidic or dry soil.

Control: Hawkweed is difficult to eradicate. Maintain dense turf and avoid bare spots to prevent it from getting started. Fertilize soil as recommended.

Chickweed (Stellaria media)

Characteristics: A low-growing annual. It has a slender stem and tiny white flowers in small terminal clusters. Seeds germinate from late fall to early spring. Spreads by seed or underground runners.

Indicator of: Chickweed grows in many conditions, but it thrives in shady, moist areas.

Control: Dig out the plants and dispose of them completely, because seeds will continue to mature even after the parent plant dies. Chickweed is difficult to eradicate, so add it to the list of weeds you can tolerate.

Veronica or Speedwell (Veronica serpyllifolia and Veronica arvensis)

Characteristics: A small creeping plant that forms mats. Cup-shaped flowers are pale blue with darker stripes. Heart-shaped seed pods. Seeds germinate in spring.

Indicator of: Thin turf, and shady areas with cool, moist soil.

Control: In the spring, rake vigorously to raise the runners and stems, then mow them off and remove them from the ground.

Thistle (Cirsium species)

Characteristics: A biennial with a low rosette of prickly leaves that overwinter the first year. Large purple flower heads appear the second year.

Control: Deep taproots and creeping underground stems make thistles difficult to eradicate. Dig out as much of the taproot as possible. Reseed the area after removing. Fertilize the soil as recommended.

Moss

Characteristics: A low emerald-green plant that forms mats on the ground surface. Mosses lack true roots and are often anchored to the ground just like thin filament.

Control: Moss grows where it is too shady and moist for grass. Unless you are willing to work hard to improve the drainage of the soil and increase the amount of sunlight the area receives, you might as well add moss to the list of plants you can tolerate. You can also introduce other shade-tolerant ground covers.

Crab Grass (Digitaria sanguinalis)

Characteristics: A grass with short, light green, hairy leaves on prostrate stems. Seeds germinate from spring through fall. Crabgrass is generally not a serious problem in Duluth.

Control: Use Safe Lawn techniques to improve the success of more desirable grass varieties or plant a different ground cover.

Ants

Control: Ants like warm soil. To encourage the ants to relocate, pour lemon juice into the anthill. To eliminate anthills, you must kill the queen ant. To do this, quickly shovel the top layer of the hill and immediately pour boiling water into the nest. If ants are still present after a week, repeat this process.

2

Native Grass Lawns

People choose to substitute native grass lawns for conventional turfgrass when they want part of their garden to have a carpeted look or be used for walking or running activities, but they don't want to embrace the regimen of chores that conventional lawns require. They may also be concerned about excessive water use, health hazards of chemicals in pesticides and fertilizers, and environmental effects of fossil fuel-powered lawn tools.

A native grass lawn may fit your needs and give extra benefits over traditional turfgrass. Installing a native grass area requires work and money up front, but is expected to repay the gardener with a healthier lawn that needs little or no water, uses no pesticides or fertilizers, and needs less frequent or no mowing.

Native grass lawns take different forms depending on their projected use and the gardener's personal taste.

Grass looks like traditional turf--clipped short and remaining green through the growing season.

In some situations, turf can be an effective use of land—to handle foot traffic or play activities, if you'd rather control weeds with a mower than hand-weed or use chemicals, to control populations of mice and bugs, as a fire break near a building, or as a green expanse to open up the view. Some native grasses and sedges "give you the opportunity to have a lawnlike planting of indigenous species while eliminating the need to apply fertilizer regularly, mow every week, and use herbicides," writes noted author Stevie Daniels in the introduction to Easy Lawns, the Brooklyn Botanical Garden's slim introduction to native grass lawns. (Daniels wrote the more detailed primer The Wild Lawn Handbook, now out-of-print but still available in some used bookstores.)

Buffalo Grass: According to Daniels, buffalograss (Buchloe dactyloides) makes a fine short turf for much of the United States. Buffalo grass is a warm-season, sod-forming grass that needs full sun and grows well from Minnesota to Montana and south to Mexico. It grows best in sunny, dry areas.

Water-Wise Gardening author Thomas Christopher advises giving buffalo grass one to one-and-a-half inches of water monthly during its growing season to ensure that it stays green rather than going dormant. He cautions that over-watering any grass compacts the soil and encourages grass to develop shallow roots, making it less drought-tolerant. Over-watered lawns (such as those with automated sprinkler systems that disregard the weather) usually have roots less than four inches deep, while under-watered grasses, depending on the species, can send roots down to five feet below the surface, which helps them find their own water.

Golf course managers are experts at creating healthy turf, and economics have prompted many to investigate grasses that thrive with little water. Some golf courses have dramatically reduced their water use by replacing traditional turfgrass with native grasses. In a recent article, the director of the United States Golf Association's Green Section estimates that replanting fairways and roughs with buffalo grass could cut water use by half on mid-western golf courses.

Most broad leaf herbicides damage or destroy buffalo grass," cautions Colorado landscape architect Jim Knopf in a chapter of Easy Lawns, but weeds can be controlled by mowing a buffalo grass lawn every two to four weeks during the growing season.

The University of Nebraska extension service has developed a useful fact sheet for buffalo grass, with a photo showing the difference between male/female and female-only turf and a summary of maintenance recommendations.

Grass is longer than turf, makes a green carpet 4 to 8 inches high.

A longer lawn can survive drought and temperature extremes better and feed itself without fertilizer. Less formal than a manicured lawn, it can provide a neutral area in a landscape but with a flowing texture and lush appearance that ultra-short turf lacks.

- *Buffalo Grass:* For this look, you can also use buffalo grass. In The Undaunted Garden, Colorado author Lauren Springer writes that her unmowed buffalo grass lawn "is too short to sway in the breeze." It stands up well to the daily traffic of neighbourhood children to and from school, and it "never gets more than four to six inches tall and is a soft-textured gray-green."

  Buffalo grass seeds produce both male and female grasses on separate plants (the technical term is dioecious), and male plants send up seed-heads on thin stalks above the leaves. Gardeners who want an uninterrupted surface of leaves will need to either mow every couple of weeks during the growing season or pay more to use only female plants. Female plants also produce seed-heads, but they are shorter than the leaves and therefore hidden from view. They are available in plugs or as sod.
- *Mixed Fescues:* Use a mix of red fescue (Festuca rubra) and hard fescue (Festuca ovina var. duriuscula), cool-season grasses 6-8 inches tall that grow best in the northern U.S. in areas with medium rainfall. Fescue species are clump-forming, but some cultivars that spread have been developed, and these will create a more solid, continuous planting. Mow once in early spring to remove dead leaves and once after flowering to remove stalks. Don't fertilize. Neil Diboll, owner of Prairie Nursery in Wisconsin, developed a "no-mow" seed mix containing six varieties of fescue. He says the average height is six inches or less, but "it can reach a height of eighteen inches if it flowers (which doesn't happen often)." Miriam Goldberger of Wild-flower Farms markets a mixed fescue eco-lawn.
- *Sedge:* For partly shaded areas, try sedges. These clumping, grasslike plants will make a loose carpet that

can be easily interplanted with bulbs and other flowering plants. However, they won't stand up to heavy foot traffic. Many are not hardy north of Zone 6; Pennsylvania sedge (Carex pensylvanica) is an exception, hardy across much of the eastern and northern U.S. and Canada. In a chapter of Easy Lawns, John Greenlee of Greenlee Nursery in Pomona, California, suggests that sedges be mowed two to three times per year at most and that chemical fertilizers, pesticide, and herbicides should not be used on sedges.

Grasses grow to their natural heights and are mixed with flowering plants in a meadow or prairie. Grasses mixed with wildflowers make a colourful carpet, taller than a traditional lawn but alive with activity. Use this type of planting to create a rewarding vista or stroll garden.

You may need to mow a few times during the first year to keep broad leaf weeds and woody species from taking over while the grasses establish. After they have established, leave seed stalks up all winter to provide food for birds, then mow or burn annually in mid-spring before the ground-nesting birds build nests. You may choose to mow or burn only one-third of the area each year to allow butterflies to reproduce.

Do not fertilize meadows or prairies, advises Nigel Colborn in The Garden Floor, because "the more fertile the soil, the more successfully the grass will compete, driving out the wildflowers."

Dry Shortgrass Meadow: In dry, hot areas, Knopf recommends planting a shortgrass meadow--a mix of buffalo grass and blue grama grass (Bouteloua gracilis) with suitably aggressive wildflowers--such as blue flax (Linum perenne), purple prairieclover (Dalea purpurea), and stiff goldenrod (Solidago rigida)--that will not be smothered by the buffalo grass. Such a mix would need a dry site to avoid being overrun by taller grass species. The flowers should be planted or seeded along with the grass so they have a chance to take hold before the grass covers the ground.

Alternately, only blue grama grass can be used as a basis for a dry meadow, and this bunch grass will allow more moderately competitive wildflowers, such as chocolate flower (Berlandiera lyrata), blanket flower (Gaillardia aristata), and Mexican hat (Ratibida columnifera) to establish.

Bulbs and Buffalo Grass: Springer doesn't recommend using buffalo grass in a meadow planting because it is "aggressive enough that perennial wildflowers have a hard time competing." However, she added spring-flowering bulbs among her buffalo grass plugs to extend the season. She writes: "... the grass had proven so vigorous and filled in so well in just three months that it was a real pain grubbing down into the mats to get to the soil to poke in the bulbs."

She has bulbs now until early June, then she rakes away the bulb foliage and dead grass from the previous year, and the lawn requires no other maintenance. "The dense turf of the buffalo grass and the lack of irrigation both see to it that weeds have a hard time gaining any foothold," she declares with satisfaction.

Fescue Meadow: Wildflowers can be added to Diboll's "no-mow" fescue mix for a short meadow.

Prairie: Diboll, whose nursery designs custom prairie seed mixtures for locations across the United States, singles out three native prairie grasses that "make excellent companions for wildflowers": little blue stem (Andropogon scoparius), sideoats grama (Bouteloua curtipendula), and Indian grass (Sorghastrum nutans). All are warm-season bunch grasses. He cautions that a Wild-flower planting without grasses will attract them, as they've evolved to fill the gaps between the flowers. For lowest maintenance, therefore, plant a mix of grasses and flowers.

Beautiful Laawns, Naturally!

Lawns without Pesticides are Beautiful: Americans care about their lawns. They care enough to spend billions of dollars to the lawn care industry annually. The problem is that in recent years, many Americans have entertained the myth that lawns must be fed a diet of chemically based fertilizers, pesticides, herbicides and fungicides in order to look presentable. Decades of advertising and public relations promotions have built the myth. Consumers pay twice to sustain this myth: first through their own pockets and then through the price of contaminated water supplies and endangered public health.

There is a ground swell of support for a change of attitudes about lawns. Many homeowners would like to discontinue the use of chemicals in their landscape but do not know how. It is becoming clear to them that the current fascination with an ideal lawn is not so ideal after all and, in fact, lawns without pesticides are more beautiful.

The Wellesley Pesticide Awareness Campaign is working to educate people about the hazards of pesticides, the benefits and beauty of organic lawns, and is helping them to transition to healthier lawn care methods. In an effort to encourage people to adopt pesticide-free practices, we are providing this booklet of information on how to manage commonly encountered lawn and yard pests.

Soil Testing

Many of us have spent the winter dreaming of the changes we'll make in our gardens and yards this season. Once the soil dries out, the first step before implementing new landscaping will be to take a soil test. A soil test allows you to add only materials that your lawn needs. Remember to request the percentage of organic matter in your sample, which should be at least 3.5 percent or more, along with organic recommendations.

Fertilizing and liming, particularly of lawns, should always follow the soil test recommendations. Don't fall into the habit of just adding a little more nitrogen or phosphorus, as over-fertilizing can cause pollution and lawn problems.

Use organic fertilizers and soil amendments, as these release their nutrients slowly and in a ratio that the plant can absorb. A balanced organic fertilizer will add the major plant nutrients: nitrogen (N), phosphorus (P) and potassium (K) along with micronutrients and trace minerals. All these amendments are important for plant growth and a healthy soil, which is alive with microorganisms and earthworms which keep it aerated and fertile. Plants growing in healthy soil develop strong roots, stems and leaves and are better able to resist disease and insect damage.

Organic fertilizers will often have an N-P-K ratio in the range of 4-3-3, a typical low analysis. If an "organic" fertilizer contains either N, P, or K near the 10 range, chances are it contains a substance prohibited in an organic programme, and you risk run-off of excess nitrogen. Study the label carefully. Avoid synthetic ingredients such as superphosphate and harmful "natural ingredients" such as nitrate of soda which contains excess salts.

Synthetic chemical fertilizers, such as 10-10-10, are highly water soluble, releasing ALL their nutrients within a short time. Excess nutrients are washed into our ground water, polluting our lakes, streams and drinking water with nitrates and phosphates. After a few weeks the synthetic fertilizer has completely dissolved and is no longer available to the plant. These petroleum-based chemical fertilizers tend to kill off beneficial microorganisms that keep the soil alive and healthy.

Organic compost is a wonderful soil amendment and the best source of minerals, nutrients and beneficial organisms. Follow soil test recommendations for the use of compost. It can be applied at any time, but spring and late summer top-dressings of 1/4" depth are preferred. Read labels carefully and avoid any product that contains sewage sludge (biosolids), which may contain harmful heavy metal contaminants. Wetting agents (synthetic polymer additives) in compost should also be avoided. The Soil Testing Laboratory at UMass Amherst can analyze compost samples using the same testing form as for soil samples.

Crabgrass

Weeds are tolerated in an organic lawn to varying degrees. Many so-called "weeds" are beneficial to the lawn ecosystem. It wasn't until the advent of selective herbicides that a lawn consisted of only grasses. Before that, any plant that lived under the mower blade was considered "lawn." This diversity of species led to lawns which were more tolerant of adverse conditions.

Often excess weeds are a symptom of poor soil, a flag that lets you know a pH adjustment is needed. Correcting the soil pH ultimately controls weeds, as healthy turf and weeds cannot co-exist. The turf will shade and crowd out the weed.

In some cases, weeds can get out of hand and take over sections of lawn. If weeds need to be controlled, there are several organic products on the market approved for use. Corn gluten meal acts as a pre-emergent weed control that suppresses the establishment of plant weeds. It is applied in the spring between Forsythia and Lilac bloom, before weeds such as crabgrass emerge. It can be reapplied in early summer to prevent germination of late-season weed seeds

Because corn gluten meal is high in protein, it contains approximately 10 percent organic nitrogen. This nitrogen level needs to be accounted for in the overall lawn management programme, so that nitrogen is not over applied during the season through additional fertilizer and compost applications, for example. The bare spots in the lawn where crabgrass was last year, treated with corn gluten, can be planted with an annual rye grass to "hold the spot" until a perennial lawn seed mixture can be established in the fall. Seed applied after corn gluten should be mixed with compost and the compost spread at least 2 inches thick in the bare areas.

Corn gluten meal is also effective in weedy flower beds, if applied before the weeds appear. Dust the meal on the bed as if it were salt and pepper hitting the dinner plate. (It doesn't matter if the corn gluten contacts foliage. It only has pre-emergent properties, not post-emergent.) During the first season of use, apply the corn gluten meal again to the bed in the fall once the weeds have died back for winter.

Unwanted weeds that exist in the lawn or border, such as plantain or dandelions, can be organically controlled with the use of non-selective herbicides made from the essences of vinegar and lemon juice (ethanoic and acetic acids) or potassium salts. The weeds are cautiously spot sprayed, being careful to avoid any unnecessary over-spray or drift onto desirable lawn or plants. Two good products for this purpose are Burn Out™ and Scythe™. Always read the label directions carefully before use.

Dandelions can also be physically removed with a two pronged metal hand tool which digs up the tap root.

Grub Control

As your lawn begins to green up, do you see bare, brown patches of soil here and there? Maybe you noticed them last fall, too. These areas are a sign that grubs (beetle larvae) have been feeding on the grasses in your lawn, literally pulling them out by their roots. Another sign of grub activity may be crows gathering on the lawn to tear up small patches or mole tunneling. Moles are voracious eaters of grubs and will quickly excavate large areas in search of their prey.

Wait-don't reach for those chemicals! This year, try an organic approach to this problem. When the soil warms up to 55 degrees, milky spore bacteria (Bacillus popilliae) may be applied for safe, long-term control of the white Japanese and oriental beetle grubs.

While it may take two to three years to achieve complete control, milky spore is worth using due to its simple application method and effective long-term results. The spores multiply inside the grubs, and when the current infestation subsides, the spores lie dormant in the soil for up to 15 years, waiting to attack subsequent grub populations.

Be sure to follow the label instructions regarding application rate. It is most important to apply the milky spore powder on an overcast day and to water it in well to the lawn and surrounding garden bed areas where beetles have been a problem.

For short-term, fairly fast-acting control of Japanese beetle grubs, try Heterorhabditis bacteriophora. This is a beneficial nematode, a microscopic soil organism, that attacks deep-dwelling lethargic white grubs. Beneficial nematodes also need to be applied when direct sun will not dry them out, and they need to be watered in well, immediately after application.

Researchers are still exploring the use of other beneficial nematodes which show promise for control of other grub species in lawns. When renovating a lawn, select insect and disease-resistant cultivars of turf grass as a preventative strategy.

These include fescues, ryes, and some bluegrasses. Nature itself often can keep grub populations in check. A dry summer can kill thousands of grubs.

Handpicking is an ideal way to keep the adult beetle population under control if there are a limited number of plants being attacked. This is easiest to do in the early morning or late afternoon when the adult beetles are more dormant - you can just push them off the leaves. Use a drop of biodegradable soap in a jar of water to collect the feeding adults. Contrary to some traditions, there is no need to use oil or kerosene in the jar. Once they have died in the soapy water, just dump the whole jar in your trash, or better yet, your compost pile

Watering

Plants need water. It's a fact of life. If you plant it, you'll have to water a new acquisition daily and then weekly for the first few months to get new roots established. Sometimes we're lucky, and Mother Nature provides rainfall here and there to help us out. Many of us have chosen to install an automatic irrigation system to supplement what nature provides. This can be both good and bad for our plantings.

As we begin to think in environmentally-friendly ways, a good rule of thumb to keep in mind is "water only when needed." An established organic lawn has an extensive root system, enabling it to find food and water even when it is dry. Such an enhanced root system will enable the lawn to survive many a dry spell.

During plant establishment, provide water as needed for the first season. Then wean the plants off water as soon as possible. A new lawn should be "syringe watered" with many light waterings until the new lawn is mowed several times. Then withhold water. The same goes for other types of plants. Water frequently to get the plant established, then water only during extended dry periods or drought.

How will you know if the plant has enough water, especially during a drought? Set up a rain gauge to measure rainfall, then plan to supplement with additional watering only as necessary. A empty tuna can tucked in an out-of-view spot makes a good home-made rain gauge to collect the one inch of water per week that is the maximum amount needed.

One inch of overhead watering (from Mother Nature or you) will effectively soak six inches down to the root zone of your lawn and plantings. You can easily check to see if the top six inches of soil is moist by using a long, narrow trowel. Avoid frequent shallow watering that does not penetrate to this six-inch depth, as this practice encourages surface root development, Japanese beetle grub infestation, fungus and root rot.

During periods of drought, a soaker hose system can be a life-saver for beds and borders. It will deliver water directly to the root zone where plants need it the most. Systems can be connected to timers to help those who travel.

With automatic overhead irrigation systems, be sure to install a dry-weather sensor that will turn on the system only during dry periods. Systems designed to run every day, rain or shine, waste a tremendous amount of water. Set the timer to run in the early morning, so that moisture can quickly dry on plant leaves. Avoid watering in the middle of the day, when evaporation loss is greatest and much water is wasted. Overhead watering should also be avoided in the evening, when cooler night temperatures inhibit drying, often leading to fungal diseases.

Mulching at the proper depth conserves moisture and is a control aid for disease, insect damage and weeds. Mulch the entire perennial, vegetable or herb bed with one to two inches of shredded leaves, grass clippings, pine needles, buckwheat or cocoa hulls. Wood chips, shredded leaves or finely-ground bark mulches are an ideal mulch at a two to four inch depth around trees and shrubs. However, do not allow the mulch to touch the bark. Finally, if you really want to save water, plant drought-tolerant species. Many perennials and herbs with silver leaves fall into this category. As an added bonus, their soft colour complements both bright, bold colours and the quieter hues.

Human Biting Insects

How wonderful it is to have the warm weather return! But there are three annoying insect pests that often can make the summer less than enjoyable: fleas, ticks and mosquitoes. Happily, there are organic controls which can be applied by the landscaper or the do-it-yourselves which can help eradicate the problem.

First, learn to properly identify the pest. There are many types of fleas, ticks, and mosquitoes. A good reference book for basic identification and information is Ninety-nine Gnats, Nits and Nibblers by May R. Berenbaum (University of Illinois Press, 1989).

Gardeners working in their yards, as well as children and pets, can pick up ticks without knowing it, as ticks lurk in vegetation. It's a good practice to conduct a tick check when returning indoors from an outdoor activity. Pay special attention to the scalp area and warm places such as armpits. Deer ticks are a major carrier of Lyme disease. Lyme disease can transmitted by ticks after they have been embedded in you for 24 hours. Tick checks performed every night before bed can greatly reduce the risks of getting Lyme disease. If you find a tick embedded on you or your pet, remove it with tweezers, then wash the area with soap and water or rubbing alcohol. Save the tick in a zip-sealed plastic bag for identification and possible testing for Lyme.

If it is a Lyme disease carrying tick, call your doctor or local health department and ask for a fact sheet on Lyme, which will describe the symptoms of the disease. Do not squeeze an engorged tick, as it may regurgitate the Lyme disease vector into your body. Use tweezers where the head enters the body.

Fleas are very annoying and cause great discomfort to pets and people. Both dogs and cats carry them. They also lurk in vegetation and can be picked up while working and playing outdoors. Look for natural flea repellents for your pets, and make sure each pet wears a non-toxic flea collar. Keep pet sleeping areas frequently washed and vacuumed. Borax dries out fleas and can be used in the lawn or bedding.

In addition, over 20 species of fleas feed on humans. Flea bites are distinctively clustered in groups of two or three, particularly around the ankles and legs. They itch persistently.

A barrier of wood chips or gravel approximately three to five feet wide surrounding the perimeter of your property will drastically reduce the flea and tick population. Why? They do not cross such a barrier. They love moisture and hate dry, uneven surfaces. Diatomaceous earth is a good repellant for ticks - it scratches and irritates their exoskeletons.

We all know the annoying whine and bite of the mosquito! With the ever-present threat of West Nile virus, which is transmitted by mosquitoes, a good preventative strategy is to rid your property of areas of standing water where mosquitoes may breed. Watering cans, tires, buckets, and even children's toys can all catch water. Empty them after each rainstorm or automatic irrigation interval. Better yet, turn these items upside down after use or take them indoors. Bird baths and water gardens without agitated water should also be avoided. Do not let water stand for more than four days. Mosquitoes also are responsible for transmitting heartworm to dogs.

Mosquito Dunks™ are donut-shaped floating larvicide dispensers containing Bt (Bacillus thuringiensis). Float these doughnut-shaped bricks in standing water for unequalled control of larvae without harm to beneficial insects. One doughnut treats 100 square feet of stagnant, standing water for 30 days. These are useful for containers such as decorative ponds which are difficult to drain on a regular basis.

Garlic sprays are the key to organic control of these three insect pests. Garlic repels rather than kills insects, so it is important to apply it before they have taken up residence, if possible. A spray can be made at home by liquefying 3 peeled cloves of garlic in the blender half filled with water. Strain out the garlic, and dilute with enough water to make a gallon. Use this to fill a spray bottle or sprayer. A drop or two of a biodegradable liquid soap will help conduct the solution through the spray mechanism more easily. Also, three natural oils that repel mosquitoes and gnats - lemon grass oil, peppermint oil, garlic oil - can be purchased as Dr. T's Mosquito and Gnat Scat™. Clay is the inert ingredient.

A new product on the market called Garlic Barrier™ can be diluted in water for those who would like the ease of using a safe, effective, ready-made mixture. Follow the label directions carefully, which suggest a dilution rate of 1 part Garlic Barrier to 10 parts water. For large areas, such as a backyard abutting a wooded area, hire a landscaper who is willing to use a backpack or large wheeled sprayer to do the job.

Repeat the spraying of the property at least once during the summer in order to keep the pests under control. Before an outdoor party, have the property sprayed two days in advance. The garlicky odor quickly dissipates minutes after application.

Poison Ivy

It's one of the worst possible itches, and most of the time, you don't even know where or how you picked it up. Poison ivy, Rhus radicans, is a native plant that is usually less than welcome in the suburban yard or garden.

While its fruits provide valuable food for birds, the oils secreted by its leaves, stems and roots are toxic to humans. Indeed, the birds are the usual suspects in spreading the plant in garden beds or at the edges of the property.

It thrives in both sunny and shady situations, in rich woodlands and in barren waste places. Poison ivy has the irritating habit of insinuating itself amidst other plants, perhaps as a survival mechanism. Often it will climb the nearest vertical structure, including trees and telephone poles. Or you may encounter it running along the ground.

Learn to recognize its "leaves of three" and "let them be" when walking in the woods. Another useful adage is, "Berries white, poisonous sight." The leaves are alternate on the stem and are divided into three leaflets.

The leaf surface may be either glossy or dull green, and either smooth or somewhat hairy. Leaf colour may vary due to the degree of shade (and weather conditions) from light to deep green, yellow-green or even red. Autumn colour is yellow and bright red before the leaves drop. A white berry-like fruit remains through winter. At all seasons of the year, it causes skin irritation in humans. Domesticated pets may also come in contact with the plants and spread the harmful oils.

What if you want to clear a neglected corner that is infested with poison ivy to make a new garden space? Or what if a few seedlings of poison ivy are starting to make a toehold in an established shrub border or perennial bed? Is there an effective, alternative way to eradicate it without using herbicides?

The answer is a resounding "Yes." Poison ivy will eventually die out if it is repeatedly clipped to the ground by mowing. In rural areas, grazing animals, such as sheep or goats, can clip it quite effectively. On suburban properties, after clipping it yourself, place a smothering mulch of cardboard topped with wood chips, bark mulch or shredded leaves over the area.

This can be a very effective way to eradicate large patches. Small seedlings may be repeatedly hand-pulled using strong, non-absorbent protective gloves. Be sure to search down to at least six inches to find the running roots. Visit the area again within two weeks to check for resprouting.

Or try the "Glove of Death" technique, a chemical-resistant glove with an absorbent glove slipped over it. Dip this glove in Burn Out™ or Sythe™ (lemon juice and vinegar solution) and wipe on the foliage. Repeat a week later, if necessary, to completely kill the poison ivy.

Vines growing up fences or trees can be cut at ground level. Several days later, pull out as much of the wilted plants as possible. Or wait until winter when they can be pulled out without damaging delicate branches and will contain fewer oils.

When pulling out poison ivy, use long, plastic, newspaper bags or impermeable protective gloves to cover hands and arms up to the shoulder. Invest in a bottle of Tecnu™, a pre- or post-handling product that helps dispel the effects of any oils you may come in contact with during the cutting or pulling operations. Also, Avon makes a pre-handling product. (Read all ingredients of these carefully as they may not be suitable for sensitive skin.) Gather all plant parts in bags and dispose of the entire collection in the trash. Wash gloves completely with Fels Naptha™ soap, an old-fashioned, yellow-coloured soap, and hang to dry. Or better yet, discard them. Under no circumstances should you burn the poison ivy or pour boiling water on it, as the poisonous oils will volatize.

Wellesley Pesticide Awareness Campaign

The long-range objective of this project is to reduce the exposure of children and adults to pesticides and pesticide breakdown products which are known or probable health hazards, to protect Wellesley's water supply, and to restore Wellesley's land use style to healthier and less toxic land care practices.

Taste Bud

Taste buds are small structures on the upper surface of the tongue, soft palate, and epiglottis that provide information about the taste of food being eaten.

The human tongue has about 10,000 taste buds.

Types of papillae: The majority of taste buds on the tongue sit on raised protrusions of the tongue surface called papillae. There are four types of papillae present in the human tongue:

- Fungiform papillae—as the name suggests, these are slightly mushroom shaped if looked at in section. These are present mostly at the apex (tip) of the tongue, as well as at the sides. Innervated by facial nerve .
- Filiform papillae—these are thin, long papillae "V"-shaped cones that don't contain taste buds but are the most numerous. These papillae are mechanical and not involved in gustation. Characterized increased keratinization.
- Foliate papillae—these are ridges and grooves towards the posterior part of the tongue found on lateral margins. Innervated by facial nerve (anterior papillae) and glossopharyngeal nerve (poasterior papillae).
- Circumvallate papillae—there are only about 3-14 of these papillae on most people, and they are present at the back of the oral part of the tongue. They are arranged in a circular-shaped row just in front of the sulcus terminalis of the tongue. They are associated with ducts of Von Ebner's glands. Innervated by the glossopharyngeal nerve.

It is known that there are five taste sensations:

- Sweet, Bitter, and Umami (now sometimes called Savoury), which work with a signal through a G-protein coupled receptor.
- Salty and Sour, which work with ion channels.

Localization of taste and the human "tongue map": Contrary to popular understanding that different tastes map to different areas of the tongue, taste qualities are found in all areas of the tongue.

The original "tongue map" was based on a mistranslation by a Harvard psychologist of a German paper that was written in 1901. Sensitivity to all tastes occurs across the whole tongue and indeed to other regions of the mouth where there are taste buds (epiglottis, soft palate).

Structure of taste buds: Each taste bud is flask-like in shape, its broad base resting on the corium, and its neck opening by an orifice, the gustatory pore, between the cells of the epithelium.

The bud is formed by two kinds of cells: supporting cells and gustatory cells.

- The supporting (sustenacular) cells are mostly arranged like the staves of a cask, and form an outer envelope for the bud. Some, however, are found in the interior of the bud between the gustatory cells.
- The gustatory (taste) cells, a chemoreceptor, occupy the central portion of the bud; they are spindle-shaped, and each possesses a large spherical nucleus near the middle of the cell.

The peripheral end of the cell terminates at the gustatory pore in a fine hair-like filament, the gustatory hair.

The central process passes toward the deep extremity of the bud, and there ends in single or bifurcated varicosities.

The nerve fibrils after losing their medullary sheaths enter the taste bud, and end in fine extremities between the gustatory cells; other nerve fibrils ramify between the supporting cells and terminate in fine extremities; these, however, are believed to be nerves of ordinary sensation and not gustatory.

Houseplant Insect Control

Houseplants are an important element in most indoor environments. We rely on them to add beauty and interest to our homes as well as to public spaces such as office buildings, restaurants, and theaters. Unfortunately, houseplants may be troubled by insects and related pests.

Sometimes houseplants are infested by pests they pick up during production. No business knowingly sells infested plants, but insects and mites and their eggs are so small they often go unnoticed. Houseplants often become infested when they're set outdoors in summer. Even if plants remain indoors, they may be attacked by tiny pests that move through screens or open windows.

Regardless of how these pests find your houseplants, they're usually manageable, often by nonchemical means—if you're alert to signs of trouble and take steps to control the pests in a timely fashion.

Prevention and Detection

Prevention is the first line of defense. Check houseplants before you buy them, then isolate them from others for two or three weeks to allow undetected problems to become obvious. It's also a good idea to check plants very carefully each fall, regardless of whether they were outdoors for the summer.

Choose plants that will thrive in the amount of light you can provide. A houseplant that's stressed from inadequate light is a more likely candidate for insect problems than one that's growing vigorously, or at least steadily.

Wash smooth-leaved houseplants regularly to prevent a build-up of dust and grime. Dust filters light before it reaches the leaf surface and can also attract and harbour insects and spider mites. It looks bad, too.

Clean large leaves and stems with a moist, soft cloth. If plants have many tiny leaves, spray them with barely lukewarm water or wrap foil over the soil, then tip each plant upside down and swish it through a tub of water to which you've added a few drops of mild liquid dish-washing liquid.

Never use a feather duster to clean your houseplants. It's too easy to unknowingly transfer tiny insects, mites, or eggs from one plant to another.

Always use sterilized potting soil; garden soil may harbour insect and disease pests. Potting soil must also drain readily. When roots sit in waterlogged soil, they're likely to rot and serve as a food source for soil scavengers that live off decaying organic matter.

Check houseplants for evidence of insects whenever you water them. You'll stand a better chance of controlling these pests before they're too numerous.

Inspect both tops and undersides of leaves, particularly any that appear speckled or mottled. This may be evidence of a pest problem, though there are other causes of discoloration. A ten-power hand magnifying lens will prove helpful when looking for tiny pests, eggs, and cast skins.

Watch for honeydew, a shiny, sticky substance secreted by aphids and scale insects. You'll find it on the upper surface of leaves, as well as on table tops and other items under the plant.

Nonchemical Control

Many houseplant insect problems can be controlled using nonchemical methods, particularly if the infestation is light.

Washing: Use a soft cotton cloth dipped in mild detergent solution (1/2 teaspoon per quart of lukewarm water) to wipe small numbers of aphids, mealybugs, scales, or mites off plants with smooth leaves. You can also dislodge a light pest infestation with a forceful spray of lukewarm water.

Handpicking: It's easy to remove and dispose of large pests such as cutworms, caterpillars, and millipedes that infest houseplants outdoors in summer. Earthworms can be driven out of the soil by setting the pot in a tub of water to saturate it. You may also remove small numbers of scale insects or mealybugs with a fingernail file or pen knife.

Sticky Traps: Sticky traps are flat cards with glue on the surface. Usually coloured yellow, they capture insects that land on them. While these traps can reduce the number of flying insect pests, you can't eliminate an insect problem by using just sticky traps. Sticky traps are most useful when used to detect the presence of white flies, fungus gnats, winged aphids, and thrips which then alerts you to take additional action.

Pruning: An insect infestation may be confined to a few leaves or branches or may be particularly bad in those locations. Eliminating the worst branches will make it easier to control pests on remaining parts of the plant. If an infestation is truly isolated, pruning may be all the control that's needed.

Give up: When a houseplant is heavily infested and badly damaged, the best course of action is to throw the plant away. Minimize your losses and avoid exposing other plants to the same pest problem. If you are reluctant to discard the plant, prune it practically to the soil. If it resprouts, watch new growth carefully for signs of infestation.

Insecticides

If you still have an insect problem after trying nonchemical methods, consider using an insecticide. Insecticides are most commonly found as aerosol or ready-to-use liquid products. Some are available in ready-to-use granular form or as liquid concentrates. They're sold at nurseries, garden centres, building supply stores and discount variety stores. These products are also available for purchase through the internet.

Selecting an insecticide can be a challenge. There are increasingly fewer on the market, which makes it difficult to find an insecticide that's very effective. There is a larger selection of insecticides available for outdoor use. Plants can be taken outside and treated (if the label permits) as outdoor ornamentals.

Read the fine print to make sure the insecticide contains the active ingredient you're looking for. Don't assume houseplant insecticides are safe for all your plants. Check each label for the list of plants the product is registered to be used on. Treating your plant with an insecticide that isn't labelled for it could result in considerable damage.

If you use a product that requires dilution with water, mix only as much as you expect to use in one day, then be sure to use it that day. Spray houseplants in a well-ventilated area.

A porch or garage works well in summer; a basement might be best in cold weather.

Never use any insecticide on a houseplant that's moisture-stressed. To avoid moisture stress, water the soil a day or two before applying insecticide. Spray your plant thoroughly, especially under the leaves, then keep it out of direct sun a full day after treatment.

Houseplant Insect Pests

Spider mites are among the most serious houseplant pests. Left untreated they can multiply rapidly, causing injury, defoliation and plant death. They're not true insects, but are more closely related to spiders and ticks.

These mites are oval shaped; yellowish or greenish in colour. They're difficult to see clearly with the naked eye, measuring only 1/50th of an inch. Magnification can reveal amber-coloured mite eggs, whitish cast skins, and black fecal specks. To verify spider mite presence, place a sheet of white paper under discolored leaves. Tap the leaves, then watch for tiny moving creatures on the paper.

Spider mites thrive in dry, warm conditions. They make their way indoors in summer and sometimes as hidden guests on Christmas trees and greenery in December.

Mites first feed on the undersides of leaves, then expand their territory as populations increase, moving from stem to stem and onto nearby plants by means of fine webbing. People can also spread them accidentally on their hands, clothing, and watering cans. Always wash your hands and any tools you've used after working with infested plants.

Spider mites damage plants by piercing leaf tissue with needle-like mouthparts, feeding on sap. Usually the first sign of spider mites is a mottled or pin-prick yellow discoloration on the undersides of leaves.

Scale Insects

Scale insects don't look like typical insects. Adults secrete a waxy shell-like covering which gives them the appearance of brown or gray bumps. They're round, oval, or oyster shell-shaped, roughly 1/16 to 1/8 inch in diameter. Tiny juveniles or "crawlers" have legs and are mobile; they're more vulnerable to pesticides than the stationary adults, but are barely visible without magnification.

Scales are usually found on plant stems and the undersides of leaves especially along mid-veins. They use needle-like mouthparts to feed on plant sap, secreting sticky honeydew as an end product of that process. Heavy feeding causes leaves to yellow and drop, slows growth and stunts plants.

Control Options: Washing, physical removal, bifenthrin, permethrin, resmethrin, insecticidal soap, pyrethrins, disulfoton, imidacloprid, plant oil extracts (at least two to three applications sprayed once every 10 - 14 days are usually necessary). Because their waxy covers are so impervious to insecticides, add a few drops of liquid soap or detergent to help the material slide under the edges of the "shells."

Mealybugs

Mealybugs are soft-bodied insects, about 3/16 inch long; easily visible without magnification. Their bodies have white, waxy filaments protruding from the tail end, and look as though they've been dusted with flour. Mealybugs can be confused with powdery mildew, and sometimes are observed as "tiny cottony clusters" on stems and leaves.

Mealybugs are most common along veins on the undersides of leaves and at axils, where leaves join stems. Some mealybugs may also be found below the soil surface on the main stem. They pierce plant tissue with sharp mouthparts, then suck the sap, which results in yellowing, leaf drop, and poor growth.

Control Options: Washing, physical removal, disulfoton, bifenthrin, permethrin, imidacloprid resmethrin, pyrethrins (at least two to three applications sprayed once every 10-14 days are usually necessary).

Thrips

Thrips are very small (about 1/16 inch long) and slender; usually tan or dark coloured. Immature thrips are white, yellow, or orange. Adults can fly, jump, or run quickly. They are difficult to see without a hand lens, though they may look like little threads on the plant.

These pests may not survive long inside. However, some types of thrips are capable of living indoors on houseplants year-round. Thrips feed by scraping leaves or flowers with their rasping mouthparts, then sucking the fluid that's released. Damaged leaves develop irregular silvery streaks or splotches. Flowers become streaked or distorted. Where feeding is heavy, you may also see small shiny black drops of excrement on the leaves.

Control Options: Washing, bifenthrin, permethrin, resmethrin, pyrethrins, insecticidal soap, neem oil, plant oil extracts (at least two applications sprayed once every 5 days are usually necessary).

Springtails

Springtails are very small (about 1/16 inch long), slender white or gray insects. They're commonly seen in the soil or in saucers beneath potted houseplants. Springtails jump when they're disturbed, so they're particularly noticeable right after you water. Springtails are scavengers, feeding on decaying roots and fungi. They prefer to live in damp potting soil, especially a mix containing a high percentage of peat. Often associated with houseplants that are kept quite moist, they rarely, if ever, damage the plants. They're just annoying.

Fungus Gnats

Adult fungus gnats are small (1/16 - 1/8 inch long) insects you may see flying around houseplants or resting nearby where they frequently are mistaken for fruit flies. The legless, worm-like larvae live in damp soil and are scavengers with habits similar to springtails, causing little or no damage.

Control Options: Water thoroughly, but let soil dry as much as possible without letting plants wilt, Bacillus thuringiensis H-14.

White flies

White flies are not true flies. They're more closely related to scales, mealybugs, and aphids. These pests are more commonly associated with plants growing in greenhouses than those growing in homes.

Adults are very small (about 1/16 inch long), white moth-like insects. They're easily disturbed, fluttering up when you water or handle a plant they're clinging to. Imature white flies are even smaller. They're flat, oval, scale-like insects normally found on the undersides of leaves. It's not easy to spot these pale insects.

All stages of white flies feed on plant sap, using their piercing-sucking mouthparts. Infested leaves can yellow and drop, reducing plant vigour, but usually not killing the plant.

Control Options: Washing, bifenthrin, permethrin, resmethrin, pyrethrins, neem oils, plant oil extracts (at least three applications sprayed once every 5 days are usually necessary).

Aphids

Aphids are easy to see but they are not generally common on houseplants. They're small (1/16 - 1/8 inch long) with pear-shaped soft-bodies, conspicuous legs, and antennae. Aphids found on houseplants are usually green, but could be black, brown, gray, yellow, red, or purple.

Look for aphids clustered on stems just below flower buds or newly opening leaf buds, as well as on flowers and the undersides of leaves. Aphids pierce plant tissue, suck sap, and excrete sticky honeydew similar to that of scale insects.

Control Options: Washing, bifenthrin, permethrin, resmethrin, insecticidal soap, pyrethrins, neem oil, plant oil extracts, disulfoton, imidacloprid.

Cyclamen Mites

Cyclamen mites are not true insects but are more closely related to spiders and ticks. They inhabit protected portions of plants, especially young tender leaves, buds, and flowers, but are not as common as other pests previously described.

These mites are extremely small and semi-transparent; impossible to see without magnification. They're spread by leaf to leaf contact and on hands and clothing. Infested leaves are stunted, brittle and often hairy; flower buds may be deformed and streaked. Sometimes injured leaves, buds and flowers may turn black. Stems of infested ivy plants are often leafless or have small hairy leaves. African violets typically have small, cupped hairy leaves in their centres.

Spider Mites and their Control

Spider mites are not insects but are more closely related to spiders. These arachnids have four pairs of legs, no antennae and a single, oval body region. Most spider mites have the ability to produce a fine silk webbing. Spider mites are very tiny, being less than 1/50 inch (0.4mm) long when adults.

Southern Red Mite: Many species of spider mites can be found in Ohio landscapes. The two spotted spider mite, Tetranychus urticae (Koch), and spruce spider mite, Oligonychus ununguis (Jacobi), are the most common pests. Other species with fewer host plants include: European red mite, Panonychus ulmi (Koch), found on apple trees; honeylocust spider mite, Platytetranychus multidigitali (Ewing); southern red mite, Oligonychus ilicis (McGregor), on a variety of plants; boxwood spider mite, Eurytetranychus buxi (Garman); and the oak mite, Oligonychus bicolor (Banks).

Types of Damage

Spider mites have tiny mouthparts modified for piercing individual plant cells and removing the contents. This results in tiny yellow or white speckles. When many of these feeding spots occur near each other, the foliage takes on a yellow or bronzed cast. Once the foliage of a plant becomes bronzed, it often drops prematurely.

Heavily infested plants may be discolored, stunted or even killed. Web producing spider mites may coat the foliage with the fine silk which collects dust and looks dirty. Life Cycles and Habits: Spider mite species seem to be warm weather or cool weather active pests. The twospotted, European red, honeylocust, and oak spider mites do best in dry, hot summer weather. The spruce and southern red spider mites do best in cool spring and fall weather.

All spider mites go through the same stages of development. Adult females usually lay eggs on their host plants. The eggs hatch in days to weeks into the first stage, called a larva. Larvae are round bodied and have only three pairs of legs. The larvae feed for a few days, seek a sheltered spot to rest and then molt into the first nymphal stage.

The first nymph now has four pairs of legs. The first nymphs feed a few days, rest and molt into the second nymph. The second nymphs feed, rest and molt into the adult stage. The males are usually the size of the second nymph and have pointed abdomens. The females have rounded abdomens and are the largest mites present.

Most spider mites spend the winter in the egg stage but the twospotted spider mite over-winters as adult females resting in protected places. Twospotted Spider Mite : The twospotted spider mite is an example of a 'warm season' mite. This pest has been reported from over 180 host plants including field crops, ornamental plants, house plants and weeds.

The females overwinter in the soil or on host plants. The females become active in April and May when they seek out the undersides of leaves on suitable hosts. Each female may lay over 100 eggs. A single generation may require as much as 20 to as few as five days, depending on the temperature. These mites prefer hot, dry weather and often do not reach damaging populations in cool, rainy periods.

In the summer, the adults and nymphs are white with two greenish spots. However, over-wintering females usually turn reddish-orange and can be mistaken for other mite species.

Spruce Spider Mite : The spruce spider mite is a common 'cool season' mite. This pest can be found on all types of conifers from spruces and pines to junipers and arborvitae. This mite spends the winter in the egg stage attached to host plants. The eggs hatch in March to April and the mites can complete development in 3 to 4 weeks. If summer temperatures are constantly over 90 F, this mite becomes dormant and lays resting eggs. These eggs and adults resume activity in the fall when cooler temperatures return.

Conifers often react slowly to the feeding of this mite. Yellowing and bronzing of the needles may not become apparent until the heat of the summer, even though the damage may have occurred the previous fall and spring.

Control Strategies

Early detection of spider mites, before damage is noticed, is important. The tiny spider mites can be detected by taking a piece of white paper or cardboard and striking some plant foliage on it. The mites can be seen walking slowly on the paper. If 10 or more mites per sample are common, controls may be needed.

Option-1

Cultural Control: Syringing Since rainy weather seems to knock off spider mites, using a forceful jet of water from a hose (syringing) can perform the same task. A regular syringing can keep spider mites under control on most ornamental plants in the landscape. This technique also helps conserve natural predators.

Option-2

Cultural Control: Quarantine and Inspection The twospotted spider mite is often introduced on infested bedding and house plants. When purchasing new plants, carefully inspect the lower leaf surface for any signs of mite activity. New house plants should be quarantined from other plants until you are sure that no mites are present.

Option-3

Biological Control: Predators There are numerous insects (lacewings and lady beetles) that prey on spider mites. However, the most commonly sold predators are other types of mites. Predatory mites (usually Phytoseiulus spp., Amblyseius spp. or Metaseiulus spp.) can be purchased and released onto infested plants.

Be sure to check listings to determine which species is appropriate. Some species are host specific and each predator works better under different weather conditions. If predators are used, do not apply pesticides that will kill them.

Option-4

Chemical Control: "Soft Pesticides" Most spider mites can be controlled with insecticidal oils and soaps. The oils, both horticultural oil and dormant oil, can be used. Horticultural oils can be used on perennial and woody ornamentals during the summer at the 1 to 2 percent rate. Higher rates of horticultural oil (3 to 4 percent) or dormant oil are useful for killing mite eggs and dormant adults in the fall and spring. The insecticidal soaps are useful in the warm season. Remember that mites are very tiny and soaps and oils work by contact only. Therefore, thorough coverage of the plant is necessary for good control.

Option-5

Chemical Control: Miticides Spider mites are usually not killed by regular insecticides, so be sure to check the pesticide label to see if "miticide" is present. Pesticides claiming "for mite suppression" are usually weak miticides and will not perform well. There are few products available to the homeowner. Dicofol (=Kelthane) is registered for over-the-counter use but is difficult to find. Acephate (=Orthene), dimethoate (=Cygon), chlorpyrifos (=Dursban), diazinon, disulfoton (=Di-syston), and malathion have over-the-counter product labels but are considered weak miticides.

Avermectin (=Avid), bifenthrin (=Talstar), dienochlor (=Pentac), fenbutatin-oxide (=Vendex), fluvalinate (=Mavrik), oxamyl (=Vydate), oxydemeton-methyl (Metasystox-R), oxythioquinox (Morestan), and propargite (=Omite) are restricted use pesticides available only to licensed applicators.

Additives for Cut Flowers

Additives for cut flowers are substances added to water, intended to extend the viability of the flowers. Once flowers are removed from the plant they continue to grow slowly but have a diminished capability of receiving the nutrients that are vital for their survival.

That there are different techniques of insuring that fresh cut flowers remain alive for the longest amount of time.

Sterilization: According to James C. Schmidt, a horticulturist at the University of Illinois, originally putting cut flowers in a sterilized vase is important to extending the life of the flowers. Vases can be cleaned using a household dish detergent or a combination of water and bleach.

Using these disinfectants insures that there will not be any bacteria growing within the vase that could potentially cause the plant to wilt and die at a faster rate. Schmidt also claims that cutting the flowers diagonally with a sharp knife under running water insures that they can immediately take up fresh and clean water. Re-cutting the stems periodically throughout the lifespan of the flowers will insure that there is a fresh surface from which the stems can take up water. This will allow the freshly cut flowers to last even longer.

Additives: According to the Brooklyn Botanical Garden, different additives can be used to prolong the lives of fresh cut flowers. Experiments were performed with various substances mixed with water, including aspirin, vitamin pills, vinegar, pennies, and flower food to test their effect on cut flowers' lifespans. Each plant was put into the same environment and the same type of plant was used in each vase.

This research found that the best additive for flowers was the retailer-provided "flower food" that is usually given with a bouquet. Plants are known to thrive in an environment where there are little bacteria, there is plenty of food for energy, and water uptake is encouraged. Flower foods contain an acidifier which helps to adjust the water's pH. With a lower pH the water and food conducting system within the flower can work at maximum efficiency.

The sugar in the food will be used by the plant as an energy source, which had been lost when the flower was cut away from its root. With these nutrients the plant will be able to fully develop. Finally, there are stem unpluggers that will make sure that the flower can easily take up water and nutrients that can later be used to take care of the needs of the rest of the plant. This combination gives the fresh cut flowers everything that they need to survive longer.

Adventitious: Adventitious, in botany, refers to structures that develop in an unusual place, and in medicine, it refers to conditions acquired after birth. This article discusses adventitious roots, buds and shoots, that are very common in vascular plants.

Adventitious Roots: Root developing from any part other than the radicle is called adventitious root. Adventitious roots develop on stems, leaves and even old roots. The radicle or primary root and its lateral roots are the only nonadventitious roots. Many aerial stems naturally form aerial roots. There are two types of adventitious roots. Aerial stems often have preformed root initials at nodes that will develop into roots given the proper conditions. Other adventitious roots are termed wound roots, because they do not form until the stem, root or leaf is wounded.

In some species, adventitious roots form on aerial stems, grow down into the soil and help keep the stems upright, *e.g.* prop roots of maize, screwpine and banyan. Some vine stems cling to their supports using adventitious roots, *e.g.* English ivy and philodendron. Adventitious roots may also form naturally when an aerial stem contacts the soil, *e.g.* arching branches of forsythia, blackberry and weeping willow. Some trees, such as willow, naturally shed pieces of twigs, a phenomenon known as cladoptosis. The shed twigs are often transported via rivers and streams and can form new plants if they develop adventitious roots.

Virtually the entire root system is adventitious in many vascular plants, especially ferns, fern allies and most monocots, *e.g.* grasses. In most monocot species, the radicle never develops to any extent, and adventitious roots form at the base of the stem. Such adventitious root systems are also termed fibrous root systems. Root systems of specialized underground stems (rhizomes, bulbs, tubers, corms) are all adventitious roots. The tap root systems of most dicots and gymnosperms develop from the primary root but even they often develop some adventitious roots when the old roots or lower stem is wounded. Horizontal, above ground stems, termed stolons or runners, usually develop adventitious roots at their nodes, *e.g.* strawberry.

Adventitious Buds and Shoots: Adventitious buds develop from places other than a shoot apical meristem, which occurs at the tip of a stem. They may develop on stems, roots or leaves. Shoot apical meristems produce one or more axillary or lateral buds at each node. When stems produce considerable secondary growth, the axillary buds may be destroyed. Adventitious buds may then develop on stems with secondary growth.

Adventitious buds are often formed after the stem is wounded or pruned. The adventitious buds help to replace lost branches. Adventitious buds and shoots also may develop on mature tree trunks when a shaded trunk is exposed to bright sunlight because surrounding trees are cut down. Redwood (Sequoia sempervirens) trees often develop many adventitious buds on their lower trunks. If the main trunk dies, a new one often sprouts from one of the adventitious buds. Small pieces of redwood trunk are sold as souvenirs termed redwood burls. They are placed in a pan of water, and the adventitious buds sprout to form shoots.

Some plants normally develop adventitious buds on their roots, which can extend quite a distance from the plant. Shoots that develop from adventitious buds on roots are termed suckers. They are a type of natural vegetative reproduction in many species, *e.g.* many grasses, quaking aspen and Canada thistle. The Pando quaking aspen grew from one trunk to 47,000 trunks via adventitious bud formation on a single root system.

Some leaves develop adventitious buds, which then form adventitious roots, *e.g.* piggyback plant (Tolmiea menziesii) and mother-of-thousands (Kalanchoe daigremontiana). The adventitious plantlets then drop off the parent plant and develop as separate clones of the parent.

Coppicing is the practice of cutting tree stems to the ground to promote rapid growth of adventitious shoots. It is traditionally used to produce poles, fence material or firewood. It is also practiced for biomass crops grown for fuel, such as poplar or willow.

Location of Origin: Adventitious roots and buds usually develop near the existing vascular tissues so they can connect to the xylem and phloem. However, the exact location varies greatly. In young stems, adventitious roots often form from parenchyma between the vascular bundles. In stems with secondary growth, adventitious roots often originate in phloem parenchyma near the vascular cambium. In stem cuttings, adventitious roots sometimes also originate in the callus cells that form at the cut surface. Leaf cuttings of the Crassula form adventitious roots in the epidermis.

Modification of Adventitious Root: Tuberous roots- These are without any definite shape. *e.g.* Sweet Potato.

Fasciculated Root: In this type the tuberous root occur in clusters at the base of the stem e.g asparagus,dahlia.

Nodulose Roots: This type of roots become swollen near the tips. e.g turmeric.

Stilt Roots: These roots arise from the first few nodes of the stem. These penetrate obliquely down in to the soil and give support to the plant. e.g maize, sugarcane.

Prop Roots: These roots give mechanical support to the aerial branches. The lateral branches grow vertically downward into the soil and acts as pillars. *e.g.* banyan.

3

Tender Bulbs, Corms and Tubers

School has started, the first football game is history, can autumn be far behind? Many gardeners are redirecting their efforts from flower production to ensuring the survival of their tender perennials. Tender perennials are not winter hardy in our part of the country, but with a little help from us, they can survive for many years. Tender perennials are those plants that need to be dug from the soil in the fall and wintered over in a frost-free location. Luckily we don't have to keep the entire plant, only the bulb, corm, or tuber are stored. New growth occurs from these structures after replanting next spring.

Popular tender perennials considered tender in Iowa include gladioli, cannas, dahlias, tuberous begonias and caladiums. Other lesser known tender perennials include the Peruvian lily also known as alstroemeria, agapanthus, freesia, and calla lilies. Survival of tender perennials requires more attention than simply digging the particular storage organ and putting it in a box in the basement. Specific storage conditions must be met to successfully store the plants through the winter.

Dig tender perennials just before or soon after a killing frost. If left until after a frost, the foliage will be killed and the storage organ will need to be dug within a few days to prevent rot-causing organisms from entering through the damaged stem. Corm producing plants such as gladioli, corn lily, and bugle lily can be stored in a similar fashion. Before digging, trim the foliage to within a few inches of the corm. Dig carefully to prevent damaging the corm. Brush off remaining soil. Allow the corms to cure for several weeks in a dry location with good air circulation. After drying, cut or twist off the foliage and discard the shriveled remains of last year's corm. A new corm is produced each year on top of the old one. Store the corms in old nylon stockings or onion bags at temperatures between 35 and 45 degrees Fahrenheit. The bulbs of Siberian lily and Mexican shell-flower can be treated the same way.

Dahlia, alstroemeria, and gloriosa lily produce tuberous roots. To store these plants over the winter, trim back the foliage of the plant to within a few inches after the first light frost. Dig carefully to avoid injury. With some soil attached, pack the roots between 2 and 3 inch layers of vermiculite, peat moss, sawdust, or wood shavings. Store at 35 to 45 degrees. Check frequently to remove those that shrivel or rot.

The tubers of tuberous begonias and elephant's ear (Colocasia esculenta) and the rhizomes of calla lilies should be dug before a hard frost. Cut the tops back and allow 2 inches of stem to remain. Dry for 2 to 3 weeks in a frost-free location, shake off the soil and remove the dried stem. Pack in peat moss, vermiculite, sawdust, or wood shavings and store at 45 to 55 degrees. Caladium (Caladium xhortulanum) tubers should be cured for a week in a warm location and stored in packing material at a temperature around 60 degrees.

Canna and butterfly lily rhizomes should be dug after the foliage has been killed by a frost. Cut the stems back to about 4 inches above the soil. Dig the rhizomes and dry in a frost-free location for about two weeks. Place the roots in shallow boxes; they do not require covering. Store at 45 to 50 degrees.

The storage organs of most tender perennials multiply quite quickly in the garden. It is important to leave them intact until spring. Any injury incurred prior to storage will increase the chances for rot to occur. In the spring cut the rhizomes and tubers apart making sure at least one or two dormant buds are present on each section. Share the extras with relatives, neighbours, and friends.

The major problem homeowners have in storing these tender perennials is finding a location with the correct temperature.

Many of us no longer have an unheated basement or extra bedroom in which to store the tender perennials adequately. Normal interior temperatures are too warm. Most garages, even though attached, will be too cold for survival. If this is your situation, it would be best to grow these plants as annuals instead of perennials.

Roots and Tuber Crops

The Root/tuber Crop Resource

Useful insights into the post-harvest loss complex of root crops are provided by a study of their characteristics. The principal roots and tubers are derived from nine species. They are all of high Moisture content, and their parenchyma cells are packed with starch grains. These materials show variable degrees of inherent keeping life, from some species of yam, like Dioscorea alata which remains dormant for 3-4 months to cassava in which there is no natural dormancy.

When such material is stored in the fresh state, respiration and transpiration continue with inevitable losses of water and dry matter. Prolonged high levels of transpiration result in a change of texture affecting quality and weight loss. The dry matter loss, caused by superation constitutes a real food value as distinct from the moisture loss which only has an economic value. Sprouting at the end of dormancy can result in dramatic loss as the physiological state is altered. At this stage the stored starch is transformed to sugars and utilized by the elongating shoots with appreciable loss of both food value and moisture.

The main root/tuber crops are cassava (Manihot esculenta Crantz), the yam (Dioscorea spp.), the potato (Solanum spp.), the sweet potato (Ipomoea batatas L. (Lam) and the edible aroids (Colocasia spp. and Xanthosama sagittifolium). With the yam, the potato, and the aroids many different species are used in the cultivated complex. Differences between these are usually sufficient to merit separate treatment.

Tare (Colocasia esculenta var esculenta) illustrates the system of maintenance of continuous food supply well. This species which has no seasonal growth constraint is planted in moist situations conducive to year-round growth. By systematic replanting whenever material is harvested, continuous supplies are assured. A similar absence of seasonality in cassava (Manihot esculenta) together with its ability to produce mature roots at various time intervals after planting permits year-round harvesting in most ecological systems. By opting for continuous food supply the opportunity for post-production losses is restricted to the food preparation phase. The same benefits accrue from the use of crop scheduling

Maintenance of food supplies ever restricted periods through storage is an essential feature of yam-based food supply systems. The success with yam storage lies in recognition of the limits to the potential for storage and in producing an environment conducive to keeping the material in sound condition. Perceptive farmers like those in Nigeria and the Ivory Coast often construct different stores to meet the separate requirements of D. alata and the different culivars in the D. rotundata/cavenesie complex. These systems work well within the normal period of yam tuber dormancy. They are so successful in West Africa that improvements in the beat indigenous practice could hardly be expected without recourse to the technologies of reduced temperature storage, the use of controlled gas environments or ionising radiation. These latter technologies are however not commonly applicable in developing countries because of cost and lack of know-how.

Considerable quantities of roots and tubers are transformed into more durable products by drying, fermentation and commination in different combinations and sequences producing a variety of materials each with distinctive characteristics. Other benefits that often accrue from transformation include the removal of toxic substances naturally present and the convenient-to-use nature of the transformed product. Although the ultimate keeping potential is constrained by environmental conditions, transformations have good prospects for long-form maintenance of root/tuber crop supply because they convert a perishable commodity into a form with stability similar to that of the cereal grains.

A feature of much of the indigenous processing technology is its high labour requirement. Because of this, food supply strategies based on transformation tend to be restricted to economic environments where labour is available.

As a result root crop transformation is widely practiced among subsistence economies whore labour inputs have zero or low cash value. Attempts are being made to establish root/tuber crop transformation commercially, but success has been variable. Radical changes in the entire system of production may be required, as has been suggested with yam (D. rotundata/cavonenais) in Nigeria for manufacture of instant pounded yam.

The Major Root/tuber Crops

Cassava (Manihot esculenta Crantz): Known variously as manioc, yuca, tapioca and mandioca, the edible portion of cassava is a starchy root which matures in harvestable state in 8-24 months according to cultivar and climate. In addition to the roots, the leaves also have potential for use as food as for instance in pert e of Central and West Africa. Cassava cultivars are often grouped into "sweet" with relatively low contorts of cyanogenic glucosides and "bitter" with high cyanogenic glucoside content although many intermediate forms exist. The "sweet" types may be eaten raw or lightly boiled without harm while the "bitter" forma require processing to remove cyanogenic glucosides.

The cassava roots have no natural dormancy and are highly susceptible to deterioration. The nature of this deterioration is as yet not fully understood, although two factors, one an indigenous physiological one and the other associated with microbiological infection, have been identified. The physiological deterioration' of which the symptoms are tissue discolouration, can commence as soon as 24 hours after harvest. Acting independently of, but usually following, the physiological changes, microorganisms, both fungi and bacteria develop in the flesh of the roots causing additional damage and loss: this microbial deterioration usually develops only 5-7 days after harvest. (Booth, 1974; Lozano, Cook and Castano, 1978).

This rapid post-harvest deterioration of cassava roots places serious constraints on their use with fresh produce trade and on the holding of buffer stocks for large-scale processing.

The short storage life of fresh cassava imposes constraints on its distribution and use. As a result, the choice of market for which production is intended becomes influenced by location. The range over which production for the fresh market extends is therefore a function of the distance from their markets and the efficiency of transport. The time of delivery rather than distance per se is the important limiting factor.

In the vicinity of Bogota, Colombia, up to an 8-hour interval from harvest to delivery is acceptable. With the current status of road a and vehicles this permits production within a 300 km radius. With more effective preservation or a faster delivery system, this range could be extended. It is claimed that the market for fresh cassava in Colombia is so structured that as quality deteriorates, the price is lowered, thereby enabling roots of decreasing quality to be bought by successively poorer income groups.

Techniques to extend the fresh life of cassava roots have been independently developed in several countries. At the Instituto de Investigaciones Tecnologicas in Colombia deterioration has been delayed by coating tubers with a film of paraffin wax (IIT 1973). At CIAT fresh tubers were shown to have improved keeping characteristics when plant stoma are out and removed but tubers are loft unharvested for up to 14 days. Tubers so treated remained sound for as long as 30 days. This technique has much to commend it, but has not yet been tested on a large scale. Other practices developed in joint CIAT/TPI projects include storage in earth silos' and storage of roots in plastic bags and in absorbent packs. These methods appear to have some utility in small and medium scale production but have not yet been proved in large-scale commercial operations. None of those techniques have as yet attained general application.

The transformation of cassava to more durable forms, frequently in forms more convenient to use than fresh roots, is partly associated with the detoxification process for the removal of hydrocyanic acid. From the crop loss point of view these products are important since they are frequently more able to avoid the rapid deterioration of the fresh tubers. Cassava products may be classified in a variety of ways, but for simplicity these may be regarded as food starch ant whole root products. These latter may be fermented or unfermented and may either be "flour" or "gari" types.

Cassava starch is extracted from roots after pealing, washing, reaping, squeezing settling and drying. Two types of starch are produced, sour or fermented and sweet or unfermented. Sweet starch is dried immediately following extraction while sour stanch is stored for 3-4 weeks when some fermentation involving lactobacilli and yeasts occurs. Sour starch is favoured for baking and commands a better price than sweet starch at least in some countries. This latter has similar conformation to corn starch with which it competes disadvantageously.

Gari, the most commonly used form of cassava in West Africa, accounts for some 70% of the entire cassava production in Nigeria. It is estimated that between 4 and 5 million tone of roots are used each year for this purpose. Although there are close parallels between the production of gari and of Brazilian "farinha de mandioca", the West African variant has some distinctive features.

The effect of the longer fermentation affects the taste by the production of lactic acid, as wall as well as reducing the content of free hydrocyanic acid. Production at levels of indigenous technology is highly labour-intensive, but this constraint is being progressively removed through a variety of innovations aimed at increasing labour productivity. The hygroscopic nature of gari is a major constraint to its keeping quality. 1a a humid atmosphere it can absorb sufficient moisture to make it vulnerable to the growth of fungal organisms. The recent practice of packaging gari dried to a safe moisture contort of around 12% in sealed polyethylene bags enables the product to be kept in good condition for extended periods.

Farinha de mandioca is the Brazilian product from which gari derives. The stages of preparation are similar to those for gari, the roots being washed, peeled, grated, pressed, toasted and classified by size. Unlike gari, the mashed or grated mass is not permitted to ferment. The heating dries the product and allows some gelatinization and dextrinization of the starch as also occurs in gari making.

Other durable cassava products include cassava flour, dried chips and "cossettes' from which flour can subsequently be prepared by milling or grinding.

Tam (Dioscorea Spp.): With yams, storage life is determined by dormancy since attempts to store yams after sprouting has commenced are impractical. During the period of dormancy the metabolic processes continue at reduced rate thereby keeping dry netter losses at relatively low levels. The various yam species have different periods of dormancy with 3-6 months for D. rotundata and D. alata and 1-2 months with D. cavenensis or D. trifida. Pathological factors are also of great importance in causing post-harvest losses in yams.

Traditionally, in West Africa, yams are stored in "yam barns" where good but uncontrolled conditions for storage exist by building the barns in the deepest available shade. This prevents heating by the nun and also helps to maintain high humidity. Several other indigenous yam storage systems are used but moat of these are less satisfactory than the "barn" or "claie verticale".

It is generally believed that these systems are well adept d to the yam varieties and to the climatic conditions prevailing in West Africa and offer little scope for improvement at current levels of technology.

Of the methods used to extend storage only the use of ionising radiation has been an unreserved technical success. The practical application of this technique is not yet generally available. Reduced temperature storage is limited by possibility of chilling damage around 10-12 C and by the growth of fungal organisms at temperatures around 14° C. Treatment with chemicals used to suppress sprouting in other crops has had little success but active research is being pursued by TPI. Controlled gas storage has not thoroughly been investigated.

Transformation of yams to more durable forma is possible but these methods have not yet met with much commercial success. Traditional products such as "amala" are variable. This may be related to the intrusion of undesirable dark colours into the final product from the use of a mixture of varieties and the failure to deactivate the polyphenol oxidases during processing rather than to any inherent basic defect of the process.

The attempts of commercial organizations to produce an instant pounded yam failed more as a result of nontechnical problem such as the logistics of raw material supply than to technical defects in the process. New processes recently developed at the University of Ife, Nigeria, may provide the required answer. A yam flake product developed to pilot project level in the West Indies has not continued long in production at a commercial level.

The Potato (Solanum Spp.): The potato, Solanum tuberosum, originally of highland tropical origin, but developed mainly as a temperate crop, is now being developed as a lowland tropical crop. Selections tolerant to high temperature and short days are being made in several parts of the tropics. At higher elevations in the tropics it is already established as a crop of some importance.

She extension of potato growing to these areas is a major concern of International Potato Centre (Centro Internacional de Papa - CIP). This Centre works together with existing bodies selecting areas for activity on the basis of comparative advantages. This approach forms a rational basis for co-operation and avoids the sometimes unintended, but nonetheless real, tendancy towards replacement of national effort with that of an international organization.

The work on post-harvest lose in this Centre is integrated with that of the social science unit permitting a broader view of the problem. This approach, taking account of social, economic and ecological factors, could provide useful guidelines in the development of other work on post-production losses. An account of the studies in the Mantaro Valley is given by way of an illustration of the CIP approach.

The Mantaro Valley lying some 300 km to the South-East of Lima has been described by Mayer (1979). It is a riverine valley acme 60 km in length and from 2 to 22 ho in breath. Three broad ecological zones are recognized. The valley floor is some 3,200 - 3,450 m above sea level. The eastern slopes between 3,400 and 3,950 m and the western slopes of similar elevation.

Finally a high zone above 3,900 m. The flat riverine soils of the valley floor have a high productivity with a capacity to support a diversified agriculture and a high population density. In this zone a contrast between large and small farm a is noticed. In the zone of sloping land above 3,450 m soil productivity declines as dose the number of crop species. The practices on the East and West slopes differ. On eastern slopes potatoes maintain their place in cropping systems but play a diminished role on western slopes.

Under the more severe conditions of the High Zone potatoes become the most important component of the cropping system occupying some 57% of the arable land. These zones are constrained by three main environmental factors: precipitation, temperature and water balance. These are all subject to changes relative to elevation. Generally precipitation and exposure to frost increases with increasing elevation.

Several products are made from potatoes including chuno, papa seca, tungush and potato starch. Ecological zone and economic circumstances result in different comparative advantages between choice of product. The most common product at low elevation is papa seca, a parboiled dried product which requires, fuel for boiling but which utilized what would otherwise be a waste product.

This restricts production to the lower zones where fuel supplies are more assured. Hybrid or native varieties are used for papa seca since this process would not be sufficient to remove the glycoalkaloids present in the bitter varieties. Processing into chuno or tongosh which eliminates the bitter principle requires no cooking. The process however requires heavy night frosts and open areas on which to expose the potatoes to natural freeze drying. These conditions are all readily met in the high blue. Locally produced potato starch competes with commercially produced corn starch. It is clear from this example that changes due to the technological superiority of one process over another is unlikely since a different technique is adapted to meet different complexes of factors.

Potato Storage

Potatoes are stored for use as seed or for food or sale. Different methods of storage have been described by CIP (Werge, 1977) from which the following summary has been taken. Ho recognizes three main systems: house' out-building and field. Peruvian rural houses according to Stein (1961) appear to have been cosigned with storage of agricultural produce in view. The convenience and security of storage in the house is evident.

The actual technique of storage adopted appears related to the itended use for the potatoes. Potatoes to be boiled in their skins are kept is the attic where the conditions permit drying and some shrivelling, causing the potatoes to become sugary and thus more favoured. Potatoes which are to be pealed before cooking may be stored on the ground where they love leas moisture and HO remain firm.

Seed potatoes may be left on the ground on a thin layer of straw or eucalyptus leaves, in piles against the walls of houses or in shallow bins or "trojas". Where animals may have access to them if they are loft on the ground, the potatoes may be placed on a platform or chaclanka of eucalyptus branches in the rafters and covered lightly with et raw.

Field storage is more frequently adopted at higher evaluations. Here strawlined clamps, covered in turn with straw and soil, may be used. The cool moist conditions reduce dehydration but the system is vulnerable to flooding which may result in dramatic crop losses. At elevations of about 3,000 m ventilation with air at ambient temperature assists the storage process.

An interesting series of low-cost potato stores have been developed by the storage unit at CIP. These are developed and adapted to take advantage of prevailing conditions to facilitate storing. Control over the storage environment is attempted by using natural ventilation systems coupled with adequate insulation and shading. Where appropriate humidity modification using simple measures is also being researched. In the case of seed tubers, these are stored in simple low-cost stores exposed to natural diffused light which, to some extent, replaces the need for controlled low temperature storage.

Sweet Potato (Ipomoea batatas (L.)

Sweet potato is a crop of considerable unrealized potential. The roots are widely used as a carbohydrate food and recently attention has also been drawn to the nutritional value of its leaves as food.

The ability of the roots to become suberized at high temperature and humidity in the classical curing process, gives the crop potential for extended storage. Storage is particularly successful when kept at relatively cool temperatures, although, it held below 12°C, chilling damage can occur. The most successful storage occurs in the southern states of America and in Japan, and successful storage for extended periods has still to be achieved in the Tropics.

There is interest in transforming the roots to more durable forms. In Trinidad, at the University of the West Indies, a sweet-potato-based flour was shown to have utility in wheat flour substitution in bakery products. There is interest there and elsewhere in the preparation of breakfast foods from sweet potato tubers. The manufacture of dried sweet potato flakes is already an established industry in the United States.

Taro (Colocasia esculenta) and Other Edible Avoids

The edible corms of these crops are normally eaten fresh after boiling or baking. Estimates in Fiji showed high peeling and trimming losses during preparation as food.

The fresh corms to not normally remain sound for long periods, and signs of deterioration become apparent a few days after harvest. Observations in Fiji indicated that packaging in plastic bags enhances keeping quality. In the South Pacific it is traditional to keep taro by transforming it into a fermented product of which "Poi" is the Hawaiian variant. It is of interest to note that the presence of oxalic acid raphides in the corms acts as a deterrent to rats.

Bulbs, Corms and Tubers—Varieties

Acidanthera

Aconites (Eranthis)

Height 10cm (4 inches)

The bulbs should be grown in fertile, humus-rich soil that does not dry out in summer. Plant 7cm (3 inches) deep and 10cm (4 inches) apart in full sun or partial shade. Flowers late winter to early spring. Divide overcrowded clumps in the autumn.

Adoni

Height 30cm (12 inches)

Japanese wild flowers, here in unusual forms. Plant the bareroots just below the surface, 30cm (12 inches) apart, into moist, neutral to acidic soil, which has had plenty of well rotted compost or manure incorporated. Provide a cool position in full shade. Avoid disturbing plants once they are established. Flowers late winter to spring.

Alliums (Ornamental Garlic)

Height range 25-100cm (10-40 inches) (depending on variety)

Plant the bulbs approximately 20cm (8 inches) deep in any moderately fertile, well-drained soil. The smaller varieties can be closely planted for impact, however allow at least 20cm (8 inches) for larger bulbs. The leaves should be allowed to die back and not removed until completely brown. Lift and divide overcrowded clumps in late summer. Prefers full sun or partial shade. Most Alliums flower from late spring to late summer and are perfect for flower arrangements.

Anemone Blanda, De Caen, St. Brigid & Nemorosa Cultivars

Height range 10-30cm (4-12 inches) (depending on variety)

The shorter lived poppy flowered types are a favourite for flower arrangers (Remember, cut the flowers do not pull them).

Both types prefer a warm, well-drained, humus rich soil in sun or partial shade. De Caen types will arrive as tubers, soak them in water overnight before planting 5cm (2 inches) deep, 10cm (4 inches) apart. Anemone blanda and nemorosa types are suitable in areas of dappled shade and are often sold as potted plants. Water well before planting out 23cm (9 inches) apart or if corms, plant as soon as received.

Arisaema triphyllum (Jack in the Pulpit)

Height 60cm (24 inches)

Plant tubers 15cm (6 inches) deep, 15cm (6 inches) apart in moist, humus-rich, neutral to acid, well-drained soil. Prefers a cool position in partial shade. Feed with a balanced liquid fertiliser monthly when in growth. Mulch with bracken or fresh compost for the winter months. During dormant periods, keep tubers just moist but never allow to dry out completely. Flowers spring to early summer.

Arum italicum

Height 30cm (12 inches)

A decorative border plant with mottled foliage, which withers away in spring to allow for white spathe-like flowers to appear, which are followed by stalks of vivid red berries. Plant tubers about 10cm (4 inches) deep and 15cm (6 inches) apart, in well-drained, humus-rich soil, ideally in dappled shade and with shelter from winds. Flowers spring.

Autumn Crocus

Begonia

Height depends on variety (some are trailing)

Start the Begonias into growth as soon as possible, planting them 5cm (2 inches) apart in trays or individually in 7cm (3 inches) pots. Plant with hollow side uppermost, into good free-draining compost. Plant so that the tip of the corm is just level with the compost surface. Place in a cool position, with good light and ventilation, ideally a cool greenhouse or windowsill.' As they shoot, grow on at a warm temperature, protecting from strong sunlight until they become established.

Harden plants off by moving outside after all risk of frost has passed. Water regularly throughout the summer, avoiding water on the foliage as this causes scorching, and feed approximately every two weeks. Remove spent flowers regularly to prolong the flowering season. Prior to the first frost, stop watering and lay plants on their side. The foliage will die back naturally. When the foliage turns yellow, remove from the corm or tuber, also remove compost from around the corm or tuber, store in dry peat in a frost-free position. Restart your corm or tuber into growth the following March. Flowers summer.

Bessera (Coral Drops)

Height 60cm (24 inches)

A striking bulb for a sunny position outdoors in well-drained soil. Plant bulbs 5cm (2 inches) deep, and at least 5cm (2 inches) apart. Plant in full light and water freely whilst in growth. Provide warm, dry conditions during winter months. Alternatively, bulbs can be grown in containers under glass in either a conservatory or greenhouse. Half hardy. Flowers late summer to autumn.

Bletilla

Height 60cm (24 inches)

A hardy, easy to grow orchid for borders. Plant the pseudobulbs (orchid bulbs) 5cm (2 inches) deep, 45cm (18 inches) apart in moist, well-drained soil with plenty of well-rotted compost or manure incorporated. Provide a sheltered site with partial shade in summer. Mulch to ensure added protection over the winter months. Flowers spring to early summer.

Bloomeria Crocea

Height 30cm (12 inches)

Large golden spheres of golden-yellow blooms for the border or rock garden. Plant corms 8cm (3 inches) deep and 8cm (3 inches) apart, in light fertile, sandy soil in full sun or partial shade. Water well when in growth, but keep dry in summer after flowering. Flowers late spring to early summer.

Bluebells (Hyacinthoides non-scripta)

Height 20cm (8 inches)

The bulbs should be planted 7cm (3 inches) deep and 5cm (2 inches) apart. Thrives in soil with organic matter in a cool and part shaded position. Flowers spring. Divide overcrowded clumps in the autumn.

Brodiaea

Height 50cm (20 inches)

Brodiaea should be grown in a sheltered site, in light free-draining soil. If soil is heavy then bulbs can be grown in containers. Excellent as a cut flower, blooming from June to July.

Calla Lily

Canna Lilies

Height range 1.5-2.2m (5-7 feet)

Cannas will give the best display if potted prior to planting in the garden. On arrival, plant individually into 13cm (5 inches) pots, using a soil-less potting mix. Grow in a position of full light with shade from hot sun. Water freely and apply a phosphate-rich liquid fertiliser monthly. Plant into the garden mid-June after all other bedding has been planted so make sure you leave an appropriate position in your display.

After the first frost cut foliage to the ground, lift root with some soil and store in a cool dry position. When dry, store in slightly moist peat in a frost-free position over the winter months. Repeat planting out process again the following March. Flowers mid summer onwards.

Calochortus (Butterfly Tulip)

Height 45-60cm (18-24 inches)

These bulbs can be grown outdoors, 15cm (6 inches) apart, in an open site in well-drained, soil. Plant the bulbs 10-15cm (4-6 inches) deep and protect from rain during dormant period and in winter by covering with a cloche. Alternatively, bulbs can be grown in containers under glass in either a conservatory or greenhouse. Plant into a loam based compost with added grit. Place in full light and water freely whilst in growth. Provide warm, dry conditions during winter months. Each bulb will produce up to 10 flowers at intervals of one to two weeks. Flowers late spring to early summer.

Camassia

Height 90cm (36 inches)

Plant the bulbs approximately 10cm (4 inches) deep in any humus-rich, well-drained soil in full sun or part shade, also excellent for naturalising in grassland. The leaves should be allowed to die back after flowering and not removed until completely brown. Do not allow soil to become waterlogged, and mulch in winter if there is risk of frost. Flowers late spring.

Chionodoxa (Glory of the Snow)

Height 10-20cm (4-8 inches)

These small bulbs should be planted in well-drained soil in full sun, at a depth of 5cm (2 inches) and 5cm (2 inches) apart. Flowers early spring.

Cardiocrinum (Yunnan Lily)

Height 1.5-4m (5-12 feet)

An imposing beauty for the back of the border, can be grown outdoors, 45cm (18 inches) apart, in a sunny position in well-drained soil. Plant so that bulbs are just below the surface in moist, humus-rich, well-drained soil.

Water freely during dry periods, although do not allow to become waterlogged, Cardiocrinums do not enjoy hot and dry conditions either. Apply a balanced fertiliser every month when in growth in order to develop the offsets. Top dress annually and provide a leaf mulch for the winter months. Bulbs may take a few years until they bloom, and after flowering strong offsets will take over in growth. Flowers summer.

Colchicum (Autumn Crocus)

Height 20cm (8 inches)

Autumn Crocus are not related to the true crocus but the similar shaped flower has given them their common name. Plant 10cm (4 inches) deep and 25cm (10 inches) apart as soon as the bulbs are received. They thrive in a well-drained, moist, humus rich soil in sun or partial shade. The autumn flowers soon open into a blaze of colour followed in the spring by their large leaves.

Cosmos atrosanguineus (Chocolate Cosmos)

Height 75cm (30 inches)

Plant 45cm (18 inches) Apart : A delicious plant for the border or containers. Plant as soon as possible, if not keep moist until planting. Plant the tubers 10cm (4 inches) deep, either one per 30cm (12 inches) container, or into beds or borders. Provide full sun and a moderately fertile, moist and humus-rich, well-drained soil. Feed with high potash fertiliser every two weeks to assist flowering. Mulch plants during the winter with peat or bracken and in more frost-prone areas lift the tubers and store, frost-free, in damp peat until replanting the following spring. Flowers late summer.

Crocosmia (Montbretia)

Height up to 1.2m (4 feet)

A diverse range of colourful plants, easy to grow and long lasting in the border. Plant in full sun or part-shade, in soil that is moist, but not waterlogged in winter. Plant corms 10cm (4 inches) deep, 8cm (3 inches) apart. In early spring to mid summer, give your plants a good mulch, and protect from slugs and snails as necessary. Planting out into containers can also prove very decorative. Flowers summer.

Crocus

Height 8-10cm (3-4 inches)

A much-loved bulb which is suitable for the border, rockeries, growing in containers, or for naturalising. The familiar wine glass-shaped flowers bloom early and are often planted with snowdrops to welcome in the spring with a wonderful show of colour. Plant in any well-drained garden soil in sun or light shade. The bulbs should be planted 7cm (3 inches) deep and 10cm (4 inches) apart. Divide overcrowded clumps in the autumn.

Curcuma roscoeana

Height 90cm (36 inches)

A Malaysian Ginger with orange blooms. Plant 1 tuber per 20-25cm (8-10 inches) pot, provide a soil-based potting mix and plant 5cm (2 inches) deep. Provide a position in bright, filtered light indoors. Water moderately until in full growth, then increase watering and feed every month with balanced liquid fertiliser and spray plants regularly. After flowering, keep dry until new growth begins in spring. Flowers summer. Greenhouse Perennial.

Cyclamen Neopolitanum (Hederofolium)

Height 10-15cm (4-6 inches)

Most people think of Cyclamen as houseplants, but you can grow these miniature hardy versions in the garden. They are most happy in a well-drained, humus rich soil in partial shade.'Plant the corms with the indentation facing upwards. They should be approximately 5cm (2 inches) deep and 15cm (6 inches) apart. Flowers summer to autumn.

Daffodils

Dahlia

Height 1.2-2m (4-6 feet)

Garden favourites, plant the tubers directly into the garden in April. Prepare the soil well and plant 15cm (6 inches) deep and 45-60cm (18-24 inches) apart. Watch for slug damage as the new foliage appears. Protect your Dahlias from wind damage. After the first frost, lift tubers and lay the roots and foliage in a frost-free shed or garage. Once the foliage has died back, cut off approximately 15cm (6 inches) from the tuber. Store in dry peat in a frost-free position. Re-plant into the garden the following spring. Flowers summer.

Dichelostemma

Height 45cm (18 inches)

Plant 15cm (6 inches) deep in humus-rich, well drained soil. Grows well in full sun or partial shade. Protect plants with mulch of bracken, twigs or well-rotted compost over the winter months. Proves fully hardy if planted by a sunny, sheltered wall. Flowers spring to summer.

Dracunculus (Dragon Arum)

Height 75cm (30 inches)

A haunting specimen, plant 10cm (4 inches) and 60cm (24 inches) apart, deep in moderately fertile, well-drained soil. Prefers full sun or partial shade. After flowering, keep bulbs dry and warm, either by leaving in the ground or by lifting and storing. Flowers early summer.

Eremerus (Foxtail Lily)

Height 2m (6 feet)

Robust perennials with lanced foliage and an array of plumes in a bright range of colours, perfect for the border. Plant tubers 90cm (36 inches) apart and 5cm (2 inches) deep in fertile, sandy, well-drained soil in full sun, and with shelter from wind and adequate support. Mulch plants with fresh compost in autumn, and protect young growth with a dry mulch as it emerges. Flowers early summer onwards.

Eranthis

Erythronium (Dogs Tooth Violets)

Height 25-30cm (10-12 inches)

This easy to grow bulb is suitable for humus-rich, well-drained soil that does not dry out, in full sun or part shade. Plant 10cm (4 inches) deep and at a spacing of 10cm (4 inches). Flowers spring.

Eucomis (Pineapple Lily)

Height 60cm (24 inches)

An exotic bulb for borders or containers. Hardy in southern regions, or when grown against a sunny, sheltered border at the base of a warm wall. Alternatively, can be grown indoors, in containers or lifted each autumn for frost-free, overwinter storage. Grow in fertile, well-drained soil in full sun, plant bulbs 15cm (6 inches) deep, and 20cm (8 inches) apart. When growing in containers, provide a soil-based potting mix with added grit and plant bulbs just below the surface. Water freely during growth, keep dry over winter. Flowers late summer.

Freesia

Height 40cm (16 inches)

Bulbs are often especially heat treated to guarantee flowering. Pot on arrival, planting 5 corms to a 12cm (5 inches) pot or 15 corms to a 30cm (12 inches) pot. Always use a good compost. Your corms can also be planted directly into the garden, planting 7cm (3 inches) deep and 10cm (4 inches) apart. If grown indoors or out, your plants will benefit from support to ensure straight stems - small canes and string are best. Ideal as a cut flower. Flowers spring onwards.

Fritillaria

Height range 15-60cm (6-24 inches) (depending on variety)

Plant the bulbs to within four times their own depth, in fertile, well-drained soil in a sunny border or rock garden. The leaves should be allowed to die back and not removed until completely brown. Alternatively, bulbs can be grown in containers under glass in either a conservatory or greenhouse. Place in full light and water freely whilst in growth. To encourage flowering in following years apply a dressing of sulphate or potash in spring and autumn.

Galtonia (Summer Hyacinth)

Height 1.2m (4 feet)

The flowers of this gem are fragrant and borne on slender stems above fleshy foliage. Plant bulbs 12cm (5 inches) deep and 10cm (4 inches) apart, deep in fertile and moist, but well-drained soil in full sun. Performs best in a sheltered border, lift bulbs in northern regions and give winter protection. Flowers late summer.

Gladioli

Height range 25-80cm (10-32 inches)

Grow Gladioli in fertile, well-drained soil in full sun or partial shade. Make successive plantings to extend the flowering season. Apply a high-potash feed as spikes develop to boost flowering. Also continue this feed every 10-14 days until 3 weeks after flowering. Staking is recommended for taller cultivars, and especially those, which are being grown for cut flower use. Gladioli are fully hardy, although you can lift them and store dry to replant the following year. To do this, cut down foliage six weeks after blooming and dry the corm for storing. Flowers summer.

Gloriosa superba

Height 2m (6 feet)

Most suited to growing in the greenhouse or conservatory. Plant tubers 10cm (4 inches) deep into containers at least 23-30cm (9-12 inches) in diameter. Provide a moist, soil-based potting mix with added grit. Do not water until growth begins, then continue to water and feed with balanced liquid fertiliser every two weeks. Keep tubers dry over the winter months. Pot on as plants become congested in later years. Flowers summer to autumn. Half Hardy.

Hyacinth

Height 20-30cm (8-12 inches)

Hyacinths should be planted directly outside into any reasonable garden soil during September/October, planting approximately 15cm (6 inches) deep and 20cm (8 inches) apart. They prefer lots of humus in the soil and sun or light shade. Hyacinths are also ideal for winter forcing in pots and containers. If you plant your bulbs as soon as possible after despatch, then you have a very good chance of blooms at Christmas. Plant your bulbs into good bulb fibre in bulb bowls or pots. A 13cm (5 inches) bowl should be ample for three bulbs, or 15cm (6 inches) for five.

Bulbs are always more effective if planted in odd numbers. Bury the bulbs so that their tips stand just above the compost, water well before placing into a dark cupboard or plunged into a cold frame, covered in damp peat. Keep frost free, and give the bulbs 10-14 weeks of these cool, dark conditions. This is essential for them to form a good root system. Remove from this place when shoots are about 5cm (2 inches), you can then bring them into the light for flowering, the colour should then be showing a week or so later.

Ipheion

Height 15-20cm (6-8 inches)

Easy to grow bulbs producing dainty star shaped flowers, which creates the most impact when planted in clumps or drifts. Should be planted 5cm (2 inches) deep and 5cm (2 inches) apart in a well-drained soil in a place of full sun. Ipheions will increase naturally in these conditions. April to May flowering.

Iris

Height range 10-45cm (4-18 inches) (depending on variety)

Bulbous Irises should be planted in a sunny, well-drained, light soil, preferring a chalky or limey soil. Plant the bulbs 5cm (2 inches) deep and 10cm (4 inches) apart for dwarf bulbs and 30cm (12 inches) apart for larger types. After planting, feed with a high-phosphorous fertiliser to encourage large bulbs to form. 'Bearded' Iris require a sunny position and the rhizomes should be planted horizontally, with the top surface exposed above soil level. When planting, top-dress with a low-nitrogen fertiliser, and then again in spring.

Ixia

Height 40-50cm (16-20 inches)

A colourful range of bulbs for containers or the border. Should be planted 8cm (3 inches) deep and 8cm (3 inches) apart. For pots, grow in soil-based potting mix, with added sharp sand, grow in full light and feed with high potash fertiliser every 3 weeks until the foliage dies back. Grow outdoors in fertile soil in full sun, lift bulbs and store dry and frost-free after flowering. Flowers spring to early summer.

Leucojum

Height 90cm (36 inches)

Leucojum resembles the Snowdrop but is much larger, and should be planted 8cm (3 inches) deep and 10cm (4 inches) apart. Best left undisturbed to form large clumps, perfect partners with Hostas. Prefers a damp position in sun or partial shade. Flowers late spring to early summer. Attractive as a cut flower.

Liatris (Blazing Star)

Height 75cm-90cm (30 inches to 36 inches)

Easy to grow and suitable for well-drained soil in full sun. Plant 10cm (4 inches) deep and 30cm (12 inches) apart. Feathery plumes of purple or white make ideal cut flowers. Flowers summer. Hardy perennial.

Lily

Height range 50-100cm (20-40 inches) (depending on variety)

Plant in the border or into large containers. For containers allow 5 per 40cm (16 inches) pot, in the bed or border plant 15cm (6 inches) deep and 22cm (9 inches) apart. Position in full sun, with a neutral to acid soil, although they will tolerate slightly alkaline soil. Dig the area well and fork in plenty of well-rotted compost.

For heavy soil, add grit to the planting hole and cover the bulb with sharp sand before replacing the soil. Feed Lilies regularly throughout the season with liquid fertiliser to boost performance and keep soil moist at all times. After flowering, you can leave the bulbs in containers and bring into a warm and frost-free position and reduce watering to keep the plants only just moist, or lift the bulbs and store dry in straw. Asiatic Lilies flower early summer.

Lily of the Valley (Convallaria)

Height 8-10cm (3-4 inches)

A traditional garden plant, which is perfect for naturalising and borders. The pips should be planted 2.5cm (1 inches) deep and 15cm (6 inches) apart. Divide overcrowded clumps after flowering, and top-dress with leaf mould in autumn. Indoor displays can be planted outdoors after flowering. Flowers spring.

Lysichiton

Height 1m (40 inches)

Plant into fertile, humus-rich soil that is moist for most of the time. Ideal conditions are at the edge of a pond or stream in full sun or partial shade. Plant tuber 6cm (2 inches) below the soil. Flowers spring.

Mirabilis (Marvel of Peru)

Height 60cm (24 inches)

A bushy perennial that grows from a tuber. Plants are hardy in southern regions, alternatively, they can be grown in containers or lifted each autumn for storing overwinter in frost-free conditions. Store as you would Dahlias. For planting outdoors, choose a sunny position and a well-drained soil. Plant tubers just below the surface, 30cm (12 inches) apart. Water freely and apply a balanced liquid fertiliser monthly. Flowers summer.

Mediterranean Bells

Muscari (Grape Hyacinth)

Height 10-20cm (4-8 inches).

Plant bulbs 10cm (4 inches) deep in well-drained soil in either full sun or partial shade. Space Muscari macrocarpum 15cm (6 inches) apart, whilst other types can be grown in close proximity to give best effect. After flowering, ensure bulbs have a dry summer dormancy. This may be best achieved by lifting and storing, especially in the case of Muscari macrocarpum. Flowers spring.

Narcissus / Daffodils

Height range 15-45cm (6-18 inches) (depending on variety)

These are onc of the easiest and the most popular of all spring bulbs. The flowers come in a wide variety of shapes, forms and colour. Any well-drained garden soil is suitable for daffodils and they thrive in sun or light shade. Miniature daffodils can be planted in tubs, pots or allowed to naturalise in a rockery.

The taller varieties can be planted in bold groups for a sensational wall of colour. Plant as soon as possible. Plant the bulbs approximately twice the bulbs depth. They can be closely planted for impact 7.5cm (3 inches), but if allowing the bulbs to naturalise, plant 15cm (6 inches) apart. The leaves should be allowed to die back and not removed until completely brown. Lift and divide overcrowded clumps in late summer, replanting offsets at once.

Narcissus 'Autumn Gem'

Height 30cm (12 inches)

Plant bulbs as soon as possible, to ensure September emergence. Plant approximately 3cm (1 inches) deep into either open ground or containers which should be well moistened, providing a reasonably well-drained soil.

Choose a site with some shelter from harsh winds. Avoid areas which tend to be frost pockets, best performance is achieved beneath a south facing wall. Remove spent flowers as they fade. Feed well with potash after flowering to benefit the bulbs energy store, and to ensure good performance for next year, remove the flowers as they fade, and remove the bulbs (by planting into a container if necessary) into a frost-free and light environment for the winter months and water occasionally if indoors. Flowers autumn.

Nectaroscordum (Mediterranean Bells)

Height 1.2m (4 feet)

Exotic bulbs for borders. Plant the bulbs 10-18cm (4-7 inches) deep in any moderately fertile, well-drained soil. They can be closely planted for impact, but if allowing the bulbs to naturalise, plant 20cm (8 inches) apart. Prefers full sun or partial shade. Flowers summer.

Nerine (Guernsey Lily)

Height 60cm (24 inches)

A bulbous perennial, producing flowers in clusters up to 15cm (6 inches) wide. A really valuable autumn flowering bulb for a border by a sunny wall or fence. Choose a sunny position and preferably leave undisturbed for several years, performs well when overcrowded as this improves flowering. No fresh manure should be added. Best planted at the surface, or at most, to a depth of 2.5cm (1 inch). Hardy, though will not survive very low temperatures in wet conditions. Soil which may lie wet through the winter should be lightened with grit or similar before planting. Can be covered with a mulch of straw or bracken in exposed areas. Flowers early autumn.

Ornithogalum (Star of Bethlehem)

Height 25cm (10 inches)

Plant bulbs 10cm (4 inches) deep, and 8cm (3 inches) apart in well-drained soil in either full sun or partial shade. After flowering, ensure bulbs have dry winter conditions, lifting and storing may best achieve this. Flowers spring. Alternatively, Ornithogalum dubium can be grown in containers under glass in either a conservatory or greenhouse.

Oxalis (Sorrel)

Height 8cm (3 inches)

Oxalis versicolor is most suitable for growing in containers in the alpine house, provide a potting mix of equal parts loam, leaf mould and grit. Plant 5cm (2 inches) deep and at a spacing of 10cm (4 inches). Water well in growth and apply a balanced fertiliser monthly. Keep barely moist when not in growth. Other forms of Oxalis are suitable for any well-drained soil in the garden. Provide a position in full sun or part shade. Flowers summer to winter.

Paris

Height 75cm (30 inches)

A plant for the shady border with amazing, colourful blooms. Plant 5cm (2 inches) deep, 30cm (12 inches) apart in moist, fertile and leafy soil in full or partial shade. Do not disturb plants and they will multiply and establish well. Flowers late spring to summer.

Polianthes (Tuberose)

Height 1.2m (4 feet)

Intensely fragrant, white blooms, perfect for containers or borders, will perform best if bulbs are lifted for winter storage. Plant 10cm (4 inches) deep, 15cm (6 inches) apart in containers of soil-based potting mix and feed with balanced fertiliser every two weeks, or plant into the border, providing a fertile, well-drained soil in full sun. Reduce watering after flowering and lift bulbs after the first frost and store in sand in frost-free conditions. Bulbs may be hardy in sheltered borders. Flowers summer to early autumn. Half Hardy.

Puschkinia

Height 20cm (8 inches)

Very easy to grow in well-drained soil, should be planted 5cm (2 inches) deep and 5cm (2 inches) apart. Completely hardy and very trouble free. Flowers early spring.

Ranunculus (Persian Buttercup)

Height 25cm (10 inches)

Bright plants for the container or the borders. Should be planted 'claws' downwards 7cm (3 inches) deep and 7cm (3 inches) apart in a rich soil, incorporating well rotted manure to ensure strong plants with large flowers.' Provide a border sheltered from the north, also suitable for growing in containers. Ideal as cut flowers.' Flowers summer.

Rhodohypoxis

Height 10cm (4 inches)

Small, clump-forming perennials in a range of bright colours, perfect for the rockery or front of the border, or patio containers. Plant 2.5cm (1 inches) deep and 10cm (4 inches) apart, into moderately fertile, well-drained, humus-rich soil in full sun, protect plants from excessive winter wet. If growing in containers, provide an acidic potting mix and grow in full light, applying a balanced liquid fertiliser monthly. Keep only just moist in winter. Flowers summer.

Sanguinaria

Height 15cm (6 inches)

A beautiful, pure white flower for woodland borders. Plant 30cm (12 inches) apart in humus rich, moist and well-drained soil in shade. Will tolerate a position that has sun for part of the day yet remains moist. Flowers spring.

Schizostylis (Kaffir Lily)

Height 60cm (24 inches)

A clump-forming perennial with sword-like foliage along with open, cup-shaped blooms in shades of red or pink during autumn. Plant rhizomes an inch below the surface of the soil and 23cm (9 inches) apart, providing well-drained, fertile and moist conditions. Shelter plants from cold, drying winds and ensure soil is always moist when plants are in full growth. Flowers autumn.

Scilla

Height 10-15cm (4-6 inches)

Scillas enjoy a humus rich, well-drained soil in a sunny or partially shaded position. Plant 5cm (2 inches) apart and 8cm (3 inches) deep or naturalise in drifts. Campanulata and Siberica flower late summer/early autumn. Mischtschenkoana is a dwarf bulb with silvery-blue flowers with darker blue stripes and flowers late winter to early spring.

Snowdrops (Galanthus)

Height 15cm (6 inches)

Plant in moist soil in light shade for the best results. The bulbs should be planted 10cm (4 inches) deep and 7cm (3 inches) apart. Divide overcrowded clumps immediately after flowering, replanting at once. Left untouched, Snowdrops can increase rapidly and form large clumps. Flowers very early spring.

Sparaxis (Harlequin Flower)

Height 25cm (10 inches)

Sparaxis are easily cultivated and should be planted in a light and sunny position, plant 5cm (2 inches) deep and 7cm (3 inches) apart.' Protect bulbs from severe frost by covering with straw or bracken, or grow in containers and lift and store dry over winter.' Flowers summer.

Tacca (Bat Plant)

Height 90cm (36 inches)

An unusual plant for the home. Plant into a special compost mix made up of equal parts leaf mould and coarse bark, with added slow-release fertiliser and provide a position in bright, filtered light. Water moderately throughout the year and spray plants regularly to keep humidity high. Apply a half-strength foliar fertiliser monthly. Pot on every 2 or 3 years and remove decaying rhizome growth. Can also be placed outdoors during the summer months. Flowers summer.

Tulips

Height range 10-60cm (4-24 inches) (depending on variety)

Tulips are members of the lily family, flowering in spring. They are best planted in November, to prevent tulip fire disease, in a warm sunny position in good well-drained soil, enriched with garden compost. Plant to a depth of 15cm (6 inches), or three times the bulb height. Keep moist during the growing season. Protect new Tulip shoots from slugs and snails by putting a protective barrier of coarse grit or sand around each plant. Slug pellets may also be used. Tulip plants leave a disease as they die off in late spring so it is important to remove the flower petals as they fall. Cut off the plants at ground level when they have turned yellow. In areas with wet summers, the bulbs can be lifted when the foliage turns yellow and stored in dry sand in a warm shed, until replanting time. Tulips also grow very well in containers. During winter keep the compost moist to dry and water well once the shoots appear.

Zantedeschia / Calla Lily

Height 60-120cm (24-48 inches)

Beautiful Arum style lilies for either containers or into open ground. 'Plant so that the tip of any visible shoots are just above the surface of the soil. Provide a good, free-draining soil with a good amount of humus and position in good light. During the growing season, water freely and feed with a balanced liquid fertiliser every two weeks until the flowers have faded. During winter, reduce watering and keep plants only just moist. Place containers in a cool greenhouse until you see signs of growth in the spring. For outside plants, overwinter by covering with twigs, leaves or moist peat. Flowers summer.

Root

In vascular plants, the root is that organ of a plant body that typically lies below the surface of the soil (compare with stem). However, this is not always the case, since a root can also be aerial (that is, growing above the ground) or aerating (that is, growing up above the ground or especially above water). On the other hand, a stem normally occurring below ground is not exceptional either (see rhizome). So, it is better to define root as a part of a plant body that bears no leaves, and therefore also lacks nodes. There are also important internal structural differences between stems and roots. The two major functions of roots are 1) absorption of water and inorganic nutrients and 2) anchoring the plant body to the ground. Roots also function in cytokinin synthesis, which supplies some of shoot needs. They often function in storage of food.

Root Structure

At the tip of every growing root is a conical covering of tissue called the root cap. It usually is not visible to the naked eye. It consists of undifferentiated soft tissue (parenchyma) with unthickened walls covering the apical meristem. The root cap provides mechanical protection to the meristem cells as the root advances through the soil, its cells worn away but quickly replaced by new cells generated by cell division within the meristem.

The root cap is also involved in the production of mucigel, a sticky mucilage that coats the new formed cells. These cells contain statoliths, starch grains that move in response to gravity and thus control root orientation.

The outside surface of the primary root is the epidermis. Recently produced epidermal cells absorb water from the surrounding environment and produce outgrowths called root hairs that greatly increase the cell's absorptive surface. Root-hairs are very delicate and generally short-lived, remaining functional for only a few days. However, as the root grows, new epidermal cells emerge and these form new root hairs, replacing those that die. The process by which water is absorbed into the epidermal cells from the soil is known as osmosis. For this reason, water that is saline is more difficult for most plant species to absorb.

Beneath the epidermis is the cortex, which comprises the bulk of the primary root. Its main function is storage of starch. Intercellular spaces in the cortex aerate cells for respiration. An endodermis is a thin layer of small cells forming the innermost part of the cortex and surrounding the vascular tissues deeper in the root.

The tightly packed cells of the endodermis contain a substance known as suberin in their cell walls. This suberin layer is the Casparian strip, which creates an impermeable barrier of sorts. Mineral nutrients can only move passively within root cell walls until they reach the endodermis. At that point, they must be actively transported across a cell membrane to continue further into the root. This allows the plant to accumulate mineral nutrients in the stele.

The vascular cylinder, or stele, consists of the cells inside the endodermis. The outer part, known as the pericycle, surrounds the actual vascular tissue. In monocotyledonous plants, the xylem and phloem cells are arranged in a circle around a pith or centre, whereas in dicotyledons, the xylem cells form a central "hub" with lobes, and phloem cells fill in the spaces between the lobes.

Secondary Growth

All roots have primary growth or growth in length. Roots of many vascular plants, especially dicots and gymnosperms, often undergo secondary growth, which is an increase in diameter. A vascular cambium forms in the stele to produce secondary phloem and secondary xylem. The epidermis is replaced by a periderm. As the stele increases in diameter, the cortex, pericycle and endodermis are lost. Even nonwoody roots often undergo secondary growth, including those of tomato and alfalfa.

Root Growth

Early root growth is one of the functions of the apical meristem located near the tip of the root. The meristem cells more or less continuously divide, producing more meristem, root cap cells (these sacrificed to protect the meristem), and undifferentiated root cells. The latter will become the primary tissues of the root, first undergoing elongation, a process that pushes the root tip forward in the growing medium. Gradually these cells differentiate and mature into specialized cells of the root tissues.

Roots will generally grow in any direction where the correct environment of air, mineral nutrients and water exists to meet the plant's needs. Roots will not grow in dry soil. Over time, given the right conditions, roots can crack foundations, snap water lines, and lift sidewalks. At germination, roots grow downward due to gravitropism, the growth mechanism of plants that also causes the shoot to grow upward. In some plants (such as ivy), the "root" actually clings to walls and structures.

Growth from apical meristems is known as primary growth, which encompasses all elongation. Secondary growth encompasses all growth in diameter, a major component of woody plant tissues and many nonwoody plants. For example, storage roots of sweet potato have secondary growth but are not woody. Secondary growth occurs at the lateral meristems, namely the vascular cambium and cork cambium. The former forms secondary xylem and secondary phloem, while the latter forms the periderm.

In plants with secondary growth, the vascular cambium, originating between the xylem and the phloem, forms a cylinder of tissue along the stem and root. The cambium layer forms new cells on both the inside and outside of the cambium cylinder, with those on the inside forming secondary xylem cells, and those on the outside forming secondary phloem cells. As secondary xylem accumulates, the "girth" (lateral dimensions) of the stem and root increases. As a result, tissues beyond the secondary phloem (including the epidermis and cortex, in many cases) tend to be pushed outward and are eventually "sloughed off" (shed).

At this point, the cork cambium begins to form the periderm, consisting of protective cork cells containing suberin. In roots, the cork cambium originates in the pericycle, a component of the vascular cylinder.

The vascular cambium produces new layers of secondary xylem annually. The xylem vessels are dead at maturity but are responsible for most water transport through the vascular tissue in stems and roots.

Types of Roots

A true root system consists of a primary root and secondary roots (or lateral roots).

The primary root originates in the radicle of the seedling. During its growth it rebranches to form the lateral roots. Generally, two categories are recognized:

- *The Taproot System:* the primary root is prominent and has a single, dominant axis; there is not dominant; the whole root system is fibrous and branches in all directions. Most common in monocots. The main function of the fibrous root is to anchor the plant.

Specialized Roots

The roots, or parts of roots, of many plant species have become specialized to serve adaptive purposes besides the two primary functions described in the introduction.

- Adventitious roots arise out-of-sequence from the more usual root formation of branches of a primary root, and instead originate from the stem, branches, leaves, or old woody roots. They commonly occur in monocots and pteridophytes, but also in many dicots, such as clover (Trifolium), ivy (Hedera), strawberry (Fragaria) and willow (Salix). Most aerial roots and stilt roots are adventitious. In some conifers adventitious roots can form the largest part of the root system.
- *Aerating Roots (or pneumatophores):* Roots rising above the ground, especially above water such as in some mangrove genera (Avicennia, Sonneratia). In some plants like Avicennia the erect roots have a large number of breathing pores for exchange of gases.
- *Aerial Roots:* Roots entirely above the ground, such as in ivy (Hedera) or in epiphytic orchids. They function as prop roots, as in maize or anchor roots or as the trunk in strangler.
- *Buttress Roots or Tabular Roots:* support roots for many tropical tree species.
- *Contractile Roots:* They pull bulbs or corms of monocots, such as hyacinth and lily, and some taproots, such as dandelion, deeper in the soil through expanding radially and contracting longitudinally. They have a wrinkled surface.
- *Coarse Roots:* Roots that have undergone secondary thickening and have a woody structure. These roots have some ability to absorb water and nutrients, but their

main function is transport and to provide a structure to connect the smaller diameter, fine roots to the rest of the plant.

- *Fine Roots:* Primary roots usually <2 mm diameter that have the function of water and nutrient uptake. They are often heavily branched and support mycorrhizas. These roots may be short lived, but are replaced by the plant in an ongoing process of root 'turnover'.
- *Haustorial Roots:* Roots of parasitic plants that can absorb water and nutrients from another plant, such as in mistletoe (Viscum album) and dodder.
- *Propagative Roots:* Roots that form adventitious buds that develop into aboveground shoots, termed suckers, which form new plants, as in Canada thistle, cherry and many others.
- *Proteoid Roots or Cluster Roots:* Dense clusters of rootlets of limited growth that develop under low phosphate or low iron conditions in Proteaceae and some plants from the following families Betulaceae, Casuarinaceae, Eleagnaceae, Moraceae, Fabaceae and Myricaceae.
- *Stilt Roots:* These are adventitious support roots, common among mangroves. They grow down from lateral branches, branching in the soil.
- *Storage Roots:* These roots are modified for storage of food or water, such as carrots and beets. They include some taproots and tuberous roots.
- *Structural Roots:* Large roots that have undergone considerable secondary thickening and provide mechanical support to woody plants and trees.
- *Surface Roots:* These proliferate close below the soil surface, exploiting water and easily available nutrients. Where conditions are close to optimum in the surface layers of soil, the growth of surface roots is encouraged and they commonly become the dominant roots.
- *Tuberous Roots:* A portion of a root swells for food or water storage, *e.g.* sweet potato and dahlia. A type of storage root distinct from taproot.

Rooting Depths

The distribution of vascular plant roots within soil depends on plant form, the spatial and temporal availability of water and nutrients, and the physical properties of the soil. The deepest roots are generally found in deserts and temperate coniferous forests; the shallowest in tundra, boreal forest and temperate grasslands.

The deepest observed living root, at least 60 m below the ground surface, was observed during the excavation of an open-pit mine in Arizona, USA. Some roots can grow as deep as the tree is high. The majority of roots on most plants are however found relatively close to the surface where nutrient availability and aeration are more favourable for growth. Rooting depth may be physically restricted by rock or compacted soil close below the surface, or by anaerobic soil conditions.

Root Architecture

The pattern of development of a root system is termed 'root architecture', and is important in providing a plant with a secure supply of nutrients and water as well as anchorage and support. The architecture of a root system can be considered in a similar way to above-ground architecture of a plant—*i.e.* in terms of the size, branching and distribution of the component parts. In roots, the architecture of fine roots and coarse roots can both be described by variation in topology and distribution of biomass within and between roots.

Having a balanced architecture allows fine roots to exploit soil efficiently around a plant, but the 'plastic' nature of root growth allows the plant to then concentrate its resources where nutrients and water are more easily available. A balanced coarse root architecture, with roots distributed relatively evenly around the stem base, is necessary to provide support to larger plants and trees.

Economic Importance

The term root crops refers to any edible underground plant structure, but many root crops are actually stems, such as potato tubers. Edible roots include cassava, sweet potato, beet, carrot, rutabaga, turnip, parsnip, radish, yam and horseradish. Spices obtained from roots include sassafras, angelica, sasparilla and licorice.

Sugar beet is an important source of sugar. Yam roots are a source of estrogen compounds used in birth control pills. The fish poison and insecticide rotenone is obtained from roots of Lonchocarpus spp. Important medicines from roots are ginseng, aconite, ipecac, gentian and reserpine. Several legumes that have nitrogen-fixing root nodules are used as green manure crops, which provide nitrogen fertilizer for other crops when plowed under. Specialized bald cypress roots, termed knees, are sold as souvenirs, lamp bases and carved into folk art. Native Americans used the flexible roots of white spruce for basketry.

Tree roots can heave and destroy concrete sidewalks and crush or clog buried pipes. The aerial roots of strangler fig have damaged ancient Mayan temples in Central America and the temple of Angkor Wat in Cambodia. Vegetative propagation of plants via cuttings depends on adventitious root formation. Hundreds of millions of plants are propagated via cuttings annually including chrysanthemum, poinsettia, carnation, ornamental shrubs and many houseplants.

Proteoid Root

Proteoid roots, also known as cluster roots, are plant roots that form clusters of closely spaced short lateral rootlets. They may form a two to five centimetre thick mat just beneath the leaf litter. They enhance nutrient uptake, possibly by chemically modifying the soil environment to improve nutrient solubilisation. As a result, plants with proteoid roots can grow in soil that is very low in nutrients, such as the phosphorus-deficient native soils of Australia.

They were first described by Adolf Engler in 1894, after he discovered them on plants of the family Proteaceae growing in Leipzig Botanic Gardens. In 1960, Helen Purnell examined 44 species from ten Proteaceae genera, finding proteoid roots in every genus except Persoonia; she then coined the name "proteoid roots" in reference to the plant family in which it was known to occur. Proteoid roots are now known to occur in 27 different Proteaceae genera, plus around 30 species from other families, including Betulaceae, Casuarinaceae, Eleagnaceae, Leguminosae, Moraceae and Myricaceae. Similar structures also occur in species of Cyperaceae and Restionaceae, but their physiology is yet to be studied.

Some Proteaceae, such as Banksia and Grevillea, are valued by the horticulture and floriculture industries. In cultivation, only slow-release low-phosphorus fertilizers should be used, as higher levels cause phosphorus toxicity and sometimes iron deficiency, leading to plant death. Crop management should minimise root disturbance, and weed control should be via slashing or contact herbicides. Horticulture aside, few plants with proteoid roots have any economic value. The only cultivated crop with proteoid roots is Lupinus albus (White Lupin).

Root Vegetable

Root vegetables are underground plant parts used as vegetables. They are called root vegetables for lack of a better generic term, but include both true roots such as tuberous roots and taproots, as well as non-roots such as tubers, rhizomes, corms, and bulbs. Several types contain both taproot and hypocotyl tissue, and it may be difficult to distinguish the two.

Regardless of anatomical type, root vegetables are generally storage organs, enlarged to store energy in the form of carbohydrates. They differ in the concentration and the balance between sugars, starches, and other types of carbohydrate.

Of particular economic importance are those with a high carbohydrate concentration in the form of starch. These starchy root vegetables are important staple foods, particularly in tropical regions. They overshadow the cereals throughout much of West Africa, Central Africa, and Oceania, where they are used directly or mashed to make foufou or poi.

List of root vegetables by anatomical type:

- True root
    - o Taproot (some types may incorporate substantial hypocotyl tissue)

- Apium graveolens (celeriac)
- Arctium spp. (burdock or gobo)
- Arracacia xanthorrhiza (arracacha)
- Beta vulgaris (beet and mangelwurzel)
- Brassica spp. (rutabaga and turnip)
- Bunium persicum (black cumin)
- Daucus carota (carrot)
- Lepidium meyenii (maca)
- Pachyrhizus spp. (jicama and ahipa)
- Pastinaca sativa (parsnip)
- Petroselinum spp. (parsley root)
- Raphanus sativus (daikon and radish)
- Scorzonera hispanica (black salsify)
- Sium sisarum (skirret)
- Tragopogon spp. (salsify)
    - o Tuberous root
- Conopodium majus (pignut or earthnut)
- Ipomoea batatas (sweet potato)
- Manihot esculenta (cassava)
- Mirabilis extensa (mauka or chago)
- Psoralea esculenta (breadroot, tipsin, or prairie turnip)
- Smallanthus sonchifolius (yacon)
- Modified stem
    - o Corm
- Amorphophallus konjac (konjac)
- Colocasia esculenta (taro)
- Eleocharis dulcis (Chinese water chestnut)
- Ensete spp. (enset)
- Erythronium spp. (katakuri)
- Sagittaria spp. (arrowhead or wapatoo)
- Xanthosoma spp. (malanga, cocoyam, tannia, and other names)
    - o Rhizome
- Zingiber officinale (ginger, galangal)

- Curcuma longa (turmeric)
- Panax ginseng (ginseng)
- Arthropodium spp. (rengarenga, vanilla lily, and others)
- Canna spp. (canna)
- Cordyline fruticosa (ti)
- Maranta arundinacea (arrowroot)
- Nelumbo nucifera (lotus root)
- Typha spp. (cattail or bulrush)
  - Tuber
- Apios americana (hog potato or groundnut)
- Cyperus esculentus (tigernut or chufa)
- Dioscorea spp. (yams, ube)
- Hemerocallis spp. (daylily)
- Helianthus tuberosus (jerusalem artichoke or sunchoke)
- Lathyrus tuberosus (earthnut pea)
- Oxalis tuberosa (oca or New Zealand yam)
- Solanum tuberosum (potato)
- Plectranthus edulis and P. esculentus. (kembili, dazo, and others)
- Stachys affinis (Chinese artichoke or crosne)
- Tropaeolum tuberosum (mashua or anu)
- Ullucus tuberosus (ulluco)
- Other
  - Bulb
- Allium spp. (garlic, onion, shallot, et cetera)
- Camassia quamash (quamash)
- Lilium spp. (lilies)

4

Flowering Bulbs

Tulips and daffodils are spring favourites of many Iowans. Some individuals, however, are unaware of the wide variety of hardy, spring-flowering bulbs. Little known bulbs, such as snowdrops and Siberian squill, offer exciting opportunities for gardeners. Consider some of the following spring-flowering bulbs when planting this fall.

Ornamental Onions (Allium spp.) are grown for their colourful flower clusters. Flowers are white, yellow, or pink to purple. They bloom from late spring to early summer. Alliums grow best in full sun. The Giant Onion (Allium giganteum) produces pinkish purple flowers in a dense, globe-shaped cluster 4 to 6 inches across. The solitary heads are borne atop a 3- to 4-foot-tall stem. The giant onion usually blooms in late June. Lily Leek (Allium moly) bears yellow flowers in a loose umbel in late spring. Lily leeks grow 8 to 12 inches tall and are best utilized in borders and rock gardens. Blue Globe Onion (Allium caeruleum) produces clear blue flowers on 1- to 2-inch-diameter flower clusters. The flowers are borne atop 1- to 1 1/2-foot tall stems.

Glory-of-the-Snow (Chionodoxa luciliae) is one of the earliest spring-flowering bulbs. Flowers are star-shaped, bright blue with white centres. Plant glory-of-the-snow in drifts in rock gardens, borders, and edgings. The 6-inch plants prefer sunny sites. 'Alba' and 'Pink Giant' have white and pink flowers, respectively.

Crocuses (Crocus spp.) grow well in either full sun or partial shade. Flower colours include yellow, blue, lavender, and purple. Many blossoms are striped. Plant crocuses in groups of 20 or more of the same colour for maximum effect. They are best planted in rock gardens, around trees and shrubs, and among low-growing groundcovers. (Do not plant crocuses in the lawn. The grass will have to be mowed before the plant foliage dies down.) Plants are approximately 3 to 6 inches tall. The Showy Crocus (Crocus speciosus) actually blooms in the fall.

Fritillaries (Fritillaria spp.) are an interesting group of plants with distinctive, bell-shaped flowers. The Crown Imperial (Fritillaria imperialis) is a striking, unusual 2- to 3-foot plant. Large bell-shaped flowers are clustered atop a 1- to 2-foot stalk in mid-spring. The flowers are topped by a tuft of foliage. Flowers are available in shades of yellow, orange, and red. The crown imperial possesses a rather unusual characteristic. Its flowers and bulbs have a musky, skunk-like odor.

The crown imperial does best in full sun to partial shade. Once planted, the bulb should not be disturbed. The Guinea-Hen Flower (Fritillaria meleagris) bears nodding, bell-shaped flowers in a distinctive checkered pattern in shades of purple, bronze, gray, and white. The Guinea-hen flower grows 12 to 15 inches tall and does best in full sun to light shade. The Persian Fritillary (Fritillaria persica) produces plum purple, pendant, bell-shaped flowers on 2- to 3-foot-tall flower spikes. Its foliage is grey-green.

Snowdrops (Galanthus nivalis) are among the first flowers of spring. The 6-inch plants produce white, drooping flowers. The solitary flowers are about 1 inch across. Snowdrops do best in partial to full shade. They are ideal for naturalizing under trees and shrubs or at the edge of woodlands. Snowdrops increase rapidly and soon form dense clumps.

Spanish Bluebell (Hyacinthoides hispanica) produces 12 to 15 nodding, bell-shaped flowers on 8- to 12-inch stems. Varieties are available in blue, pink, and white. Plants bloom in late spring. Spanish bluebells perform well in partial to heavy shade.

Hyacinths (Hyacinthus orientalis) are prized for their attractive, highly fragrant flowers. The flaring, out-facing flowers cover 8- to 10-inch, upright spikes. The flowers may be yellow, white, pink, red, purple, or blue. Hyacinths prefer partial to full sun and well-drained soils. They are most effective when massed in beds and borders. Hyacinths can also be forced indoors.

Dwarf Irises are bulbous species that bloom in late winter or early spring. Danford Iris (Iris danfordiae) produces bright yellow flowers with greenish brown blotches. The violet-scented flowers of the Reticulated Iris (Iris reticulata) are blue, purple, or violet with white or yellow blotches on their falls. The blossoms of Harput Irises (Iris histrioides) are light blue to blue-purple. Dwarf irises perform best in well-drained soils in full sun. They are excellent choices for rock gardens and the fronts of borders. Dwarf irises are generally 3 to 6 inches tall. Because of their small size, dwarf irises should be planted in large numbers to create an attractive display.

Spring Snowflake (Leucojum vernum) blooms in early spring. The drooping, bell-shaped flowers are white tipped with green and possess a sweet violet-like fragrance. Plants grow best in partial shade to full sun and are generally 6 to 12 inches tall. Plant masses of spring snowflakes in rock gardens or around trees and shrubs.

Grape Hyacinth (Muscari spp.) produces urn-shaped flowers (which somewhat resemble grapes) on 6- to 9-inch spikes. Flowers are various shades of blue or violet, but there are also white-flowered varieties. The Armenian Grape Hyacinth (Muscari armeniacum) is the most widely planted species because of its vigour and larger flowers. The grape hyacinths prefer full sun and are best utilized for edging beds and in naturalized areas. Grape hyacinths will often send up foliage in the fall, but will not bloom until the following spring.

Lebanon Squill (Puschkinia scilloides) blooms in early spring. Flowers are white to pale blue with dark blue stripes running down the centres of the petals. Lebanon squill performs best when planted in large masses. They multiply rapidly.

Siberian Squill (Scilla sibirica) produces bright blue flowers in early spring on 4- to 6-inch-tall plants. They are easy to grow and prefer partial shade to full sun. Siberian squill are most effective when planted in masses under trees and shrubs, but are also suited for rock gardens and the edge of woodlands. Siberian squill multiplies rapidly. 'Spring Beauty' has slightly larger flower spikes and deep blue flowers.

Bulb

A bulb is an underground vertical shoot that has modified leaves (or thickened leaf bases) that is used as food storage organs by a dormant plant.

A bulb's leaf bases generally do not support leaves, but contain food reserves to enable the plant to survive adverse conditions. The leaf bases may overlap and surround the centre of the bulb as with lilies, or may completely surround the inner regions of the bulb, as with the onion. A modified stem forms the base of the bulb, and plant growth occurs from this basal plate. Roots emerge from the underside of the base, and new stems and leaves from the upper side.

Other types of storage organs (such as corms, rhizomes, and tubers) are sometimes erroneously referred to as bulbs. The correct term for plants that form underground storage organs, including bulbs as well as tubers and corms, is geophyte. Some epiphytic orchids (family Orchidaceae) form above-ground storage organs called pseudobulbs, that superficially resemble bulbs.

Plants that form true bulbs are all monocotyledons, and include:

- Onion, garlic, and other alliums, family Alliaceae.
- Lily, tulip, and many other members of the lily family Liliaceae.
- Amaryllis, Hippeastrum, Narcissus, and several other members of the amaryllis family Amaryllidaceae.
- Two groups of Iris species, family Iridaceae: subgenus Xiphium (the "Dutch" irises) and subgenus Hermodactyloides (the miniature "rock garden" irises).

Bulbil

Some lilies form small bulb-like fruits called bulbils in their leaf axils. Several members of the onion family, Alliaceae, including Allium sativum (garlic), form bulbils in their flower heads, sometimes as the flowers fade, or even instead of the flowers. The so-called Tree onion (Allium cepa var. proliferum) forms small onions which are large enough for pickling.

Corm

A corm is a short, vertical, swollen underground stem of a plant (usually one of the monocots) that serves as a storage organ to enable the plant to survive winter or other adverse conditions such as summer drought and heat (estivation). A corm consists of one or more internodes with at least one growing point, and is typically surrounded by protective skins or tunics. Inside, a corm is mostly starch-containing parenchyma cells. Corms can be dug up and used to propagate or redistribute the plant (see, for example, taro).

Externally, they are often similar in appearance to bulbs, and erroneously called by that name, but internally their solid tissue easily distinguishes them from bulbs, which are visibly layered.

Cultivated plants that form corms include:

- Many plants of the family Iridaceae grown for their flowers, including Crocus, Gladiolus, Iris, and Montbretia
- Many plants of several families grown as root vegetables; see that article for a list.

Tuber

A tuber, or stem tuber, is a development of a stolon, or above-ground rhizome, thickened for use as a storage organ. The tuber usually, though not always, develops when the stolon becomes subterranean, such as in a potato. This modified stem forms at the end of a lateral branch. Auxiliary buds are located at the nodes. When the buds develop, adventitious roots and rhizomes form below them. This should not be confused with a tuberous root, sometimes called a root tuber, such as a sweet potato or Dahlia.

Tuber (genus)

The true truffles are a group of several valuable and highly sought-after edible species of underground ascomycetes belonging to the fungal genus Tuber. All are ectomycorrhizal and are therefore found in close association with trees.

The ascoma (fruiting body) of truffles is highly prized as food. It has a smell similar to deep-fried sunflower seeds or walnuts, though not all people are able to catch the odor of this fungus. The water after soaking truffles can taste a bit like soy sauce. Brillat-Savarin called the truffle "the diamond of the kitchen" and praised its aphrodisiacal powers. While the aphrodisiacal characteristics of truffles have not been established, it is still held in high esteem in traditional French, northern Italian and Istrian cooking, and in international haute cuisine.

The origin of the word truffle appears to lie in the Latin term tuber, meaning "lump" or "truffle", which early on became tufer- and gave rise to the various European terms: French Truffe, Spanish Trufa, German Trüffel and Dutch Truffel. Interestingly the Italian tartufo and Piedmontese tartifla suggest an affinity with the German Kartoffel "potato".

Biology

Strictly speaking, the true truffles are those several species of the genus Tuber; however the term has been applied to several other genera of underground fungi around the world. These include the genera Terfezia and Tirmania of the family Terfeziaceae, known as the Desert truffles of Africa and the Middle East. The term "Hart's truffle" has been used to refer to Elaphomyces while "Bohemian truffle" has been used to describe Pisolithus tinctorius, which was historically eaten in parts of Germany.

True truffles form symbiotic relationships with several varieties of deciduous trees, including oak, beech, hazel and poplar, but cannot be found on maples nor on cedars. They flourish throughout the fall, winter and spring, and can be found buried between the fallen leaves and twigs and the mineral soil. Their growth beneath the earth's surface is thought to be an adaptation to forest fires, drought, or severe cold, where the mushrooms on the surface of the soil are more prone to getting destroyed.

Black Truffle

The Black truffle or Black Perigord Truffle (Tuber melanosporum) is named for the Perigord region in France and grows exclusively with oak. Specimens can be found in late autumn and winter, reaching 7cm in diameter and weighing 100g though are usually much less. Production is almost exclusively European, with France accounting for 45%, Spain 35%, Italy 20%, and small amounts from Slovenia and Croatia. In 1900, France produced around 1,000 metric tonnes (1,100 short tons) of Tuber melanosporum.

Production has considerably diminished in the past century, and is presently around 20 metric tonnes (22 short tons) per year, with peaks at 46 metric tonnes (50 short tons) in the best years. 80% of the French production comes from southeast France: upper-Provence (departements of Vaucluse and Alpes-de-Haute-Provence), part of Dauphine (departement of Drome), and part of Languedoc (departement of Gard); 20% of the production comes from southwest France: Quercy (departement of Lot) and Perigord.

The largest truffle market in France (and probably also in the world) is at Richerenches in Vaucluse. The largest truffle market in southwest France is at Lalbenque in Quercy. These markets are busiest in the month of January when the black truffles have their highest perfume. Black truffles on these markets sell between 200 and 600€ per kilogram ($110-$330 per pound), depending on the quantity and quality of the harvest.

White Truffle

The White truffle or Alba Truffle (Tuber magnatum) hails from the Langhe area of the Piedmont region in northern Italy and, most famously, from the city of Alba. Growing symbiotically with oak, hazel, poplar and beech and fruiting in autumn, they can reach 12cm diameter and 500g, though are usually much smaller.

The flesh is pale cream or brown with white marbling. Like the French black truffles, Italian white truffles are very highly esteemed (illustration, right). The white truffle market in Alba is busiest in the months of October and November. The Tuber magnatum truffles sell between 200€ and 400€ per hectogram ($1000 - $2000 per pound). They produce a scent that mimics a male pig sex hormone, and for this reason, female pigs have been used historically in Europe to help find truffles. However, more recently, dogs have become preferred for truffle hunting since they can be trained to just find the truffles whereas sows eat the truffles as soon as they find them.

The Tuber magnatum pico White truffle is mostly found in northern and central Italy, while the Tuber borchi, or Whitish truffle, is found in Tuscany, Romagna and the Marche. Neither of these is as aromatic as those from Piedmont.

Chinese Truffle

The Chinese truffle (Tuber sinensis, also sometimes called Tuber indicum) is mass harvested in China. It resembles Tuber melanosporum, but its taste is bland, and its texture is chewy. Due to their low price, Chinese truffles are often exported to the West, but they should not be confused with Tuber melanosporum. Some truffle exporters or delicatessen shops sell Chinese truffles into which extracts of the real Tuber melanosporum are introduced.

These fraudulent truffles are sold at a high price, pretending they are real Tuber melanosporum. Such practices are illegal, but unfortunately quite frequent. Another type of Chinese truffle is the Tuber himalayensis, which visually looks so much like the Tuber melanosporum that a microscope is needed to differentiate them, but whose taste is not as intense as the Tuber melanosporum.

The Tuber himalayensis, however, is harvested in very small quantities in the Chinese Himalayas, and is not as frequently met on world markets as the Tuber sinensis. Finally, the third type of Chinese truffle is the Chinese summer white truffle, which does not have a scientific name yet. This truffle should not be confused with the much more expensive Italian Tuber magnatum.

Summer Truffle

The Black Summer Truffle (Tuber aestivum/unicinatum) thrives in northern Italy, central Europe and the UK but also grows in Turkey and North Africa. It is highly valued for its culinary uses and costs up to $670 a pound. Summer truffles don't have as strong an aroma or taste as winter truffles do. They are mainly harvested from June to November. These truffles grow in symbiosis with trees such as oaks, hazels and beech. They can weigh up to 20-30 g, and their shape is generally round up to 4 cm diameter.

Other Species

Two lesser-used truffles include the black truffle (Tuber macrosporum) and the Scorzone truffle (Tuber mesentericum). In the Pacific Northwest, several species of truffle are harvested both recreationally and commercially, most notably, the Oregon white truffles, Tuber oregonense and Tuber gibbosum.

History

Antiquity: The first mention of truffles appears in the writings of Theophrastus in the fourth century BC. In classical times their origins were a mystery which challenged many; Plutarch and others thought them the result of lightning, warmth and water in the soil, while Juvenal thought thunder and rain to be instrumental in their origin. Cicero deemed them children of the earth, while Dioscorides thought they were tuberous roots.

Italy in the Classical Period produced two kinds of truffles: the Tuber melanosporum and the Tuber magnatum. The Romans, however, only used the terfez (Terfezia bouderi), a fungus of similar appearance which the Romans called truffles, and which is sometimes called "desert truffle". Terfez used in Rome came from Lesbos, Carthage, and especially Libya, where the coastal climate was less dry in ancient times. Their substance is pale, tinged with rose. Unlike truffles, terfez have no taste of their own. The Romans used the terfez as a carrier of flavor, because the terfez have the property to absorb surrounding flavours. Indeed, Roman cuisine used a lot of spices and flavours, and terfez were perfect in that context.

Islam: It is narrated in Sahih Muslim (sayings of Muhammad) that Muhammad (P.B.U.H) said 'Truffles are (a kind of) 'Manna' which Allah (Arabic, God), the Exalted the

Majestic, sent to the people of Israel through [Moses], and its juice is a medicine for the eyes.

Terfezia was the main truffle consumed in the Middle East historically, and Ludovico di Varthema, in his Travels (1503-08), wrote of great quantities of them being sold, having been harvested in the mountains of Armenia and Turkey.

Middle Ages: Truffles were rarely used during the Middle Ages. The only trace of truffles in Medieval cooking is at the court of the popes in Avignon. Black and subterranean truffles were probably considered satanic and thus avoided, but the popes discovered them when they relocated to Avignon, near the producing regions of upper-Provence, and they became very fond of them.

Truffle hunting is mentioned by Bartolomeo Platina the papal historian in 1481, when he recorded that the sows of Notza were without equal in hunting truffles, however they should be muzzled to prevent them from eating the prize.

Renaissance and Modern Times: Truffles reappeared in Europe during the Renaissance, where they were honored at the court of King Francis I of France. However, it was not until the 17th century that Western (and in particular French) cuisine abandoned "heavy" oriental spices, and rediscovered the natural flavor of foodstuffs. Truffles were very popular in Paris markets in the 1780s, imported seasonally from truffle grounds, where peasants had long enjoyed their secret.

They were so expensive they appeared only at the dinner tables of great nobles -and kept women, Brillat-Savarin (1825) noted characteristically. The greatest delicacy was a truffled turkey. "I have wept three times in my life," Rossini admitted. "Once when my first opera failed. Once again, the first time I heard Paganini play the violin. And once when a truffled turkey fell overboard at a boating picnic."

Cultivation

Truffles long eluded techniques of domestication, as Jean-Anthelme Brillat-Savarin (1825) noted with his characteristic skepticism:

"The most learned men have sought to ascertain the secret, and fancied they discovered the seed. Their promises, however, were vain, and no planting was ever followed by a harvest. This perhaps is all right, for as one of the great values of truffles is their dearness, perhaps they would be less highly esteemed if they were cheaper.

"Rejoice, my friend," said I, "a superb lace is about to be manufactured at a very low price."

"Ah!" replied she, "think you, if it be cheap, that any one would wear it?"

However, contrary to stubborn legends, truffles can be cultivated. As early as 1808, there were successful attempts to cultivate truffles, known in French as trufficulture. People had long observed that truffles were growing among the roots of certain trees, and in 1808, Joseph Talon, from Apt (departement of Vaucluse) in southern France, had the idea to sow some acorns collected at the foot of oak trees known to host truffles in their root system.

The experiment was successful: years later, truffles were found in the soil around the newly grown oak trees. In 1847, Auguste Rousseau of Carpentras (in Vaucluse) planted 7 hectares (17 acres) of oak trees (again from acorns found on the soil around truffle-producing oak trees), and he subsequently obtained large harvests of truffles. He received a prize at the 1855 World's Fair in Paris.

These successful attempts were met with enthusiasm in southern France, which possessed the sweet limestone soils and dry hot weather that truffles need to grow. In the late 19th century, a dramatic epidemic of phylloxera destroyed much of the vineyards in southern France. Another epidemic destroyed most of the silkworms in southern France, making the fields of mulberry trees useless.

Thus, large tracts of land were set free for the cultivation of truffles. Thousands of truffle-producing trees were planted, and production reached peaks of hundreds of tonnes at the end of the 19th century. In 1890 there were 750 km^2 (185,000 acres) of truffle-producing trees. In the 20th century however, with the growing industrialization of France and the subsequent rural exodus, many of these truffle fields (champs truffiers or truffieres) returned to wilderness.

The First World War also dealt a serious blow to the French countryside, killing 20% or more of the male working force. As a consequence of all these events, newly acquired techniques of trufficulture were lost. Also, between the two world wars, the truffle fields planted in the 19th century stopped being productive. (The average life cycle of a truffle-producing tree is 30 years.)

Consequently, after 1945 the production of truffles plummeted, and the prices have skyrocketed, reaching the zenith that we know today. In 1900 truffles were used by most people, and on many occasions. Nowadays, they are a rare delicacy reserved for the rich, or used on very special occasions.

In the last 30 years, new attempts for mass production of truffles have been started. Eighty percent of the truffles now produced in France come from specially planted truffle-fields. Nonetheless, production has yet to recover its 1900s peaks. Local farmers are opposed to a return of mass production, which would decrease the price of truffles.

However, prospects for mass production are immense. It is currently estimated that the world market could absorb 50 times more truffles than France currently produces. There are now truffle-growing areas in Spain, Sweden, New Zealand, Australia, Oregon, North Carolina and the UK.

Looking for truffles in open ground is almost always carried out with specially trained pigs or dogs. Pigs were the most used in the past, but nowadays farmers prefer to use dogs. Both pigs and dogs have keen senses of smell, but while dogs must be trained to the scent of truffles, female pigs, or sows, need no training whatsoever.

This is due to a compound within the truffle similar to androstenol, the sex pheromone of boar saliva, to which the sow is keenly attracted. It may have been the strange attraction that pigs have to these fungi which prompted its discovery by early human populations. Dogs are used nowadays because they are a lot easier to manage when hunting for truffles and, unlike pigs, they will probably not eat the truffles when they find them.

In New Zealand and Australia

The first black truffles (Tuber melanosporum) to be produced in the southern hemisphere were harvested in Gisborne, New Zealand in 1993. In 1999, the first Australian truffles were harvested in Tasmania, the result of eight years of work. Trees were inoculated with the truffle fungus in the hope of creating a local truffle industry.

Their success and the value of the resulting truffles has encouraged a small industry to develop. A Western Australian venture had its 1st harvest in 2004, and in 2005 they unearthed a 1kg truffle that is potentially the largest ever harvested in the southern hemisphere. Production is expanding into the colder regions of Victoria and New South Wales.

Culinary Use

Because of their high price and their pungent taste, truffles are used sparingly. Supplies can be found commercially as unadulterated fresh produce or preserved, typically in a light brine.

White truffles are generally served uncooked and shaved over steaming buttered pasta or salads. White or black paper-thin truffle slices may be inserted into meats, under the skins of roasted fowl, in foie gras preparations, in pates, or in stuffings. Some speciality cheeses contain truffles as well.

The flavor of black truffles is far less pungent and more refined than that of their white cousins. It is reminiscent of fresh earth and mushrooms, and when fresh, their scent fills a room almost instantly. In 2006, designer Tom Ford released a perfume that lists black truffle as its first note.

Pricing

As truffles are considered a delicacy, they command high prices. The world's most expensive truffle was a 1.51 kilogram rare White Alba truffle. It was sold for 125,000 Euros (Hong Kong Dollar $1,250,000; US $160,000) on 13 November 2006 to Hong Kong property tycoon Sir Gordon Wu.

This price beat the 2005 world record of 95,000 Euros for a 1.21 kilogram White Alba truffle. The 2006 auction took place at the Castle of Grinzane Cavour in Italy, with three auction houses taking part (via satellite link): Castle of Grinzane, Hong Kong, and Paris.

Truffles are also found in the Kalahari Desert. The local people of the Kalahari call the truffle a n/abba, mahupu, sand-potato or knoll.

5

Important Features of Flowering Plants

Flowering plants (Angiospermae) represent one of the largest groups of primary producers. Their contribution to the production of oxygen as well as that to the nutriment of animals and man is consequently very large. All features reviewed in this chapter refer to seed-producing plants, also called spermatophytes.

Typically, flowering plants are organized into an underground root and a shoot above ground that consists itself of a stem and leaves. The organs of a plant that serve sexual reproduction are the flowers. Part of the pollinated flower ripens and becomes the fruit.

In contrast to many other plant groups, flowering plants are striking, numerous and common. They are the most important group of the so-called primary producers that generate the prerequisite for life on earth: oxygen. Green plants have the ability to convert solar energy into chemical energy (photosynthesis) producing the oxygen necessary for all other organisms as a by-product. The useable plants among the flowering plants are - directly or indirectly - the basis of human existence; they are, too, an important economical factor. A basic knowledge of flowering plants should therefore be among everybody's general knowledge.

Much has been written about flowering plants and every reader of this chapter will miss something that he regards worth knowing, while he might find other information trivial. But everybody will understand that it is impossible to review in a few lines a theme about which an extensive, partly popular scientific literature exists. And although this term may sometimes be used in a disparaging way, most of the popular scientific literature is scientifically correct, lucid and, above all, very well illustrated.

To get an idea of the variety of existing plants and to get to know special species, it is necessary to identify them. Many books on classification with different approaches exist. In many popular books, colour photos or drawings are used and often is the colour of the flower a primary feature of recognition. Most of the so-called scientific books on classification work on dichotomic keys, *i.e.* the user is asked a lot of questions in succession and has at each one to decide between two answers. This procedure is continued until the plant has been identified. The "scientific nature" is mostly based on the completeness of the varieties present in a book, since almost all of the illustrated books contain only the most common or most striking plants.

Recently, electronic tools have been used to figure out identification keys for plants. An example is the developed by Ray PHILLIPS, Director of Information Technology Services and Assistant Prof. of Biology, Colby College, Waterville, ME 04901.

The research into the flora of Central Europe has a tradition that goes back for centuries. It is mirrored in the nearly complete modern floras and books on classification. Incomplete, if existing at all, are books on the classification of less well discovered regions, like tropics, subtropics and many mountain areas.

The question of the origin of the wealth of forms (evolution) is discussed elsewhere where it is also shown that mountains with their rather small and isolated areas provide ideal conditions for the coming into being of new species. This is the reason, why even very experienced botanists equipped with renowned books on classification of the Central European flora will sometimes and in some places fail (Alps).

This chapter will deal with the characteristics of flowering plants that produce seed (phanerogames or spermatophytes) only. Many of the structures present in this plant group can be found with other non-flowering plants, too. But mosses, ferns and algae miss some features, like flowers or seeds while others, like roots or leaves exist in an incomplete way or are replaced by other organs.

The body of vegetation of many-celled algae (and mosses) is called thallus, that of flowering plants, ferns and fern-like plants (pteridophytes) is called cormus. The latter are therefore summed up as cormophytes. The special features of the different plant groups will be discussed later.

The body of vegetation of a "typical" flowering plant consists of an underground root and a shoot above ground. The shoot is organized into stem and leaves. Each of these basic organs can exist in many variations and these again can be combined in many different ways. The almost unlimited ability of combination is one of the main reasons of the existence of such a high number of species while at the same time, the identification of the relations of the species is aggravated.

If seemingly different organs with different functions can be traced back to the same basic organ, they are called homologous. It is also spoken of homology. Contrasting is analogy where organs with a similar look and function have descended from different organs.

The Stem

Typically, flowering plants are organized into an underground root and a shoot above ground that itself consists of a stem and leaves. The stem is the main supporting axis of a flowering plant. It is made up of nodes, the part where one or more leaves are borne and of the usually longer, in-between lying internodes. Stems are characterized by their type of branching, by their surface and their symmetry. Special cases of stems are those of climbing plants, succulence, runners and rhizomes.

Shoot and root can be clearly distinguished within a seed. After germination, they do develop into the shoot that grows towards the sun and into the root which grows towards the earth's score. The shoot that develops during the first stage of development is called the main- or the primary shoot and depending on the plant, it does bear one, two, or more cotyledons. Growing and sprouted shoots are organized into nodes and the in-between lying internodes. The nodes are the parts where the leaves are inserted. At the tip of a growing shoot sits the terminal bud of the shoot apex that is surrounded by the leaf (or flower) primordias. In the area of the nodes are axillary buds where the axillary shoots insert.

Vegetable shoots are either annual or perennial. The realised alternative depends usually on the species and in some species on the quality of the environment.

The shoot of annual species is mostly fleshy and is usually called a stalk while perennial plants lignify and thus form trunks and branches. Some species, like Graminaceae or Poaceae have a hollow stalk that is called a blade of grass.

To characterize a shoot, its type of branching and its surface (smooth, hairy, with or without grooves, round or squared) is used, as well as the ratio of heights to diameter.

It can generally be distinguished between monopodial and sympodial branching. Monopodial branching exists, if a continuous main shoot has feeble apical shoots. Sympodial branching is given, if apical shoots of different generations exist at the same time since the apical meristem of each sympodial branch has either died or is differentiated into a flower, a thorn or a tendril and has thus lost its ability to grow. The respective shoot tip is nearer to the ground than that of the next apical shoot. If only one of the apical shoots has the same direction as the main shoot would have had, then a monochasium results which may sometimes be difficult to distinguish from monopodial branching. But if two buds of a sympodial branch sprout at the same time, a more or less fork-shaped pattern, a dichasium, is formed. Hardly ever do several buds put forth at the same time (pleiochasium).

An extreme ratio of heights to diameter is found in climbing plants where the diameter (and the stability) is very small compared to the length so that upright growing is only possible with the support of other plants or growing aids. Succulence is another extreme, that is typical for plants living in arid or saline habitats.

Most shoots are of a radial symmetry but with a few species, the apical shoots may be changed into flat, leaf-like phylloclades. Often, especially with wooden plants, the axes are differentiated into long and short shoots.

The internodes of the long shoots that cause elongation with trees and shrubs are longer than the short shoots. These again are often specialized apical shoots with shortened internodes. The thorns of white-horn (Crataegus), for example, or those of sloe (Prunus spinosa) correspond to such short shoots.

Apical shoots may, too, be changed into runners or rhizomes. Runners are horizontally growing shoots with greatly prolonged internodes that are either underground or above ground and are sometimes bearing leaves.

The internodes produce sometimes roots at the part that is nearest to the main shoot (shoot-borne roots). Some parts of these runners can switch to an upright growth and thus behave like a normal shoot. They serve to propagate vegetatively (strawberries are a typical example). Rhizomes are mostly underground and grow in a horizontal (plagiotrop) manner.

They are rather enduring shoots whose internodes usually stay short. They serve as survival organs and are often at the same time used for storage; sometimes, they end in specially developed storage tubers like in potatoes. Except for small, scale-like leaves, rhizomes do usually have no leaves. Shoots which happen to come into touch with a moist soil surface may start to develop leaves and flowers.

Roots

Roots, the lower, underground part of a plant are characterized by their lack of leaves. They are the structures that function to take up water, nutriments and minerals from the soil. The primary root is the main axis of a root system. Mostly, it bears thinner side-roots. Special types of primary roots are tap roots or turnips. Adventitious roots are roots originating directly from the stem.

The most striking morphological feature of roots is their lack of leaves. Roots are organized into primary roots and side-roots. The latter may develop a frequently branched and extensive root system. Among the prominent types of roots are the tap roots where the primary root is dominant while the side-roots are hardly developed at all.

The primary root grows deep into the soil. Swollen roots that have developed into storage organs are called turnips. Their formation may involve the hypocotyl (the part between the neck of the root and the part where the leaves insert; see picture: an extremely long hypocotyl, cotyledons and primary leaves of an African leguminosa). Swollen side-roots can also become storage organs (tubers).

The root-system of many monocots (like, that of the poaceae, for example) is completely formed from adventitious roots. Adventitious roots are roots that originate directly at the shoot. Rhizomes and shoots that touch the soil are able to form adventive roots. This fact is used to propagate plants vegetatively through cuttings (parts with nodes in them).

Epiphytes do often develop air roots. They are, just like usual roots able to take up minerals and rain water and sometimes they can even perform photosynthesis.

The picture shows a two-dimensional model of a root seeking water in the soil during its development. The initial water distribution has been predetermined, forming an S-shaped zone of high concentration indicated by the light colour. The growing tips of the main root and rootlets absorb water that diffuses in the soil. The decreased water concentration is indicated by dark areas that emerge around the root system. In areas with insufficient water concentration the rootlets cease to grow before they have reached their potential full length.

A leaf is an aerial and lateral outgrowth of the stem of a usually flat and dorsiventral anatomy. It functions mainly to manufacture food by photosynthesis and consists typically of a stalk also called petiole, a flattened blade, the lamina, and the leaf base. Strands of conducting and strengthening tissues, the veins, run through it.

Their pattern, also called leaf venation, is a feature of characterization. Leaves may be simple, *i.e.* undivided or compound (composed of several parts called leaflets). The blade margin and the leaf arrangement at the stem are further features of characterization.

While potentially unlimited growth is a feature of the shoot, that of leaves is usually limited. They are almost always flat and have a dorsiventral anatomy, *i.e.* their upper and their undersurface are constructed differently and have different functions. This morphological and anatomical peculiarity is strongly connected to the main function of leaves, photosynthesis. With most conifers, the leaves are very thin and often tough (needles).

A typical leaf is organized into blade (lamina), stalk or petiole and leaf base. In some plant species, the leaf base is plainly set off against the petiole while the transition is invisible in others. Sometimes, it is merely indicated by a faint widening. Occasionally, lateral excrescences develop. Most species have clearly recognizable petioles. If they are missing, the leaves are said to be sessile in which case the leaf base may sometimes surround the shoot.

The extremes are laminas with fused basal rims evoking the impression of a leave that the shoot grows through (Claytonia perfoliata). The basis of the leaf stalk may even grow down on the shoot a little bit further. With many monocots, like Poaceae or Graminaceae, the shoot is completely surrounded by a sheath-forming petiole.

The blade is radiated by leaf veins that contain the fibres. It is distinguished between parallel, arched, pinnated or netlike vein structures.

Just as variable as the vein structure is the shape of the leaves. The fundamental difference is that between simple and compound leaves. The compound leaves are built from several small leaves or from pinnations that sit in a regular organisation at the undivided or branched rhachis.

Cataphylls are simply organized, scale-like leaves. They can be found at the sprout, at the basis of sprouting shoots or underneath the normal leaves and in wooden plants also as a protection for the buds.

Finally, bracts have to be mentioned that live above the region of the usual leaves, for example at the inflorescence as subtending bracts for the flowers. But they can also have very different functions. Remarkable are the strikingly coloured bracts of Bougainvillea spectabilis and Poinsettia pulcherrima (poinsettia) that add strongly to the attractivity of the rather inconspicuous flowers. Another example are the Cuckoo-pint and its relatives where the bract encloses the inflorescence and is at the same time a refined trap for pollinating insects (Arum maculatum).

Some species, especially those that live in water have differently shaped leaves (heterophylly). An example is Ranunculus aquatilis (water crowfoot) which has pinnation-shaped, submerse living leaves and flat swimming leaves.

Another feature of recognition is the blade margin that can be shaped very differently. It can be plain, serrated, double serrated, crenated or sinuated. Leaves with deep clefts are called pinnatified, pectinated, palmated or lobed.

Quite a few species have structures called leaf metamorphosis (variations), because they are homologous to leaves. Among these are thorns where the moulding of the blade does not occur and the leaf veins have been strengthened by the deposit of a though material, as well as leaf tendrils that also miss the lamina but have an extremely pliable and tension-proof petiole. There may, too, be succulence of leaves.

Leaves are always organized around the shoot in a regular pattern that can be recognized best when looked at from above. It is obvious in most cases that the leaves are arranged in the way of a screw and that there is always the same angle measurable between subsequent leaves (angle of divergence). In its simplest form, this angle is 180° which means that the leaves are grouped into two opposing rows along the shoot. But other angles do occur, too: 120° (a third of a circle), 144° (five leaves in two circles), or 135° (eight leaves in three circles), etc.. The regularity of the positions of leaves has been discovered in the last century by C. F. SCHIMPER and A. BRAUN. The occurring angles correspond to the FIBONACCI-series, *i.e.* numerator and denominator of following fractions can be calculated by adding the numerators and the denominators of the two preceding fractions:

2/5 = 1+1 / 2+3; 3/8 = 1+2 / 3+5 etc.

The resulting series is:

1/2 1/3 2/5 3/8 5/13 8/21 13/34 21/55 34/89 ....

Calculated down to the degree of the angles, a limit of approximately 137° 30' is achieved. This angle is known to divide an arc of a circle with the golden section.

The advantage of a regular arrangement of the leaves is in the optimal yield of light gained.

Borderline Cases

Whorl like arrangement of the leaves: At each node, two or more leaves insert. The organization of the following node usually places the leaves into the gaps of the preceding one. If two leaves spring from one node, they are without exception opposite. The leaves of the node above (and that beneath) form a right angle with those of the middle node. This is also called a decussate arrangement. Distichous means that all leaves are arranged in a vertical plane

Rosettes are usually sitting near the soil; at the basis of the shoot follows node after node with almost complete reduction of the internodes. Side shoots with greatly prolonged internodes may spring from rosettes. They have one or more flower at their tip, like, for example the primrose. Especially in biennial plants, the main shoot can, too, grow with prolonged internodes and even branch. It is not unusual that the leaves of the rosette and that of the shoot differ in shape.

Flowers

The organs of a plant serving sexual reproduction are flowers. Their final product is seed. A flower is a rather complex structure formed by the flower axis that is usually transformed into the receptacle and the perianth. The perianth consists of calyx and corolla. The calyx is made up by he sepals, the corolla by the petals, stamens and carpels. If calyx and corolla cannot be distinguished or if the calyx is missing, it is spoken of an incomplete flower or perigon. The androecium is the totality of the male reproductive organs. It comprises the stamens. Each stamen consists of a filament and an anther.

The female reproductive organs, the gynoecium, include the carpels. Carpels are structures that are made up from an ovary and a stigma and that contain one or more ovules. One or more carpels may be combined to a pistil (ovary, style, stigma), the gynoecium as a whole. Taxonomical flower features are the flower symmetry, the way in which the carpels are fused and the resulting position of the ovules, also called placentation, and the position of the ovary.

Flowers that contain both androecium and gynoecium are called androgynous or hermaphroditic. If both male and female flowers live on one plant it is called monoecious. Plants with male and female flowers borne on separate plants are termed dioecious.

Before we deal with the architectural design of flowers in more detail, it should be said that they are structures serving the sexual reproduction: their final product is seed. Flowering plants are thus also called spermatophytes (seed plants). They include two large groups, the gymnosperms and the angiosperms.

The gymnosperms are the evolutionary older and more primitive group. They are, in a strictly systematical sense, no class on their own but enclose two real classes, the conifers and the cycadeans that do not have the same ancestors. The angiosperms that are usually seen as the original flowering plants do all belong to just one class. They represent the overwhelming majority of species and what you will learn in the following is principally true for their flowers.

It is taken for granted that everybody has through his own experience gained an impression of the diversity of flower shapes, colours and smells. It will be equally well known that these features developed in an evolutionary process of mutual adjustments of flowers and their individual pollinators (insects: bees, bumble-bees, butterflies, flies, etc.; birds: hummingbirds, etc.). Plants that are pollinated via wind have very unobtrusive flowers.

Many textbooks present the flower of the anemone (Anemone nemorosa) or that of tulip (Tulipa gesneriana) as typical examples. This is no bad choice, indeed, since their basic building plan is very easily demonstrated.

Flowers are short shoots with limited growth. The leaf organs are arranged in several circles (whorls), one above the other at an often strongly compressed flower axis. These structures differ strongly both in their appearance and in their function from normal leaves.

Progressing from bottom to top and from the out- to the inside can it be distinguished between the usually unobtrusively coloured calyx that consists of sepals, the conspicuously coloured corolla that is built of petals, the stamens and finally the carpels. Calyx and corolla together are the perianth.

But if calyx and corolla cannot be distinguished or if the calyx is missing it is spoken of an incomplete flower or perigon. Flowers with a missing perianth are typical for species with wind-mediated pollination. The totality of the stamens, the male reproductive organs, is called androecium that of the female reproductive organs is named gynoecium.

The flower axis has hardly any appearance in most angiosperms. It is usually shortened and broadened to the so-called receptacle that is sometimes slightly vaulted or disc-shaped. In gymnosperms, however, as well as in some primitive angiosperms (Magnolia, for example), it has the shape of a cone.

The flowers of angiosperms have originally only had an incomplete perianth that can very likely be traced back to leaflike bracts and occurs in an even more simplified version already in gymnosperms. The majority of angiosperm flowers has a double perianth with a clear separation of function between calyx and corolla. While the corolla has as a rule the function of attracting pollinators (it is, after all, designed mainly on insect aesthetics), the calyx is made to protect the growing bud.

It loses its function as soon as the bud flowers; with some species it is even dropped off during flowering, though with most species it just stops growing or shrivels. Only rarely are the sepals reactivated after pollination. If they are, a new growth starts whereby an involucre of the fruit is formed.

All flower structures have originally been organized in a tight screw (Anemone). These flowers are strictly spoken asymmetric but are nevertheless grouped together in books on classification with those of a radial symmetry. They are characterized by the arrangement of their flowers in whorls (flowering circles). All structures of one whorl are of the same appearance and every whorl has the same number of structures (three, four and five are very common).

Subsequent whorls are usually organized in a way that brings the structures of one whorl in the gap positions of the preceding (rule of alternation). Such flowers have two or more planes of symmetry; sometimes those with only two planes are called bilateral. Even if the gynoecium shows another number of organs, the flowers are grouped together with those with radial symmetry.

Flowers that have only one plane of symmetry due to either differing shapes of the structures of one whorl or due to incomplete whorls are widespread. They are thought to be of a very high developmental state and are called zygomorph (monosymmetric or dorsiventral). Examples are the pansy (Viola) or the labiates (Lamiaceae).

With most flowering plants, the petals are separate from each other (dialypetalae) while they are fused to a tubular corolla with sympetalae. The number of free points shows how many petals are involved.

Both symmetry and architecture of the flower are clearest represented via diagrams of the flowers. They show schematized outlines where the single structures, their numbers and positions are given. If such diagrams are compared to each other, it shows that five is a typical number for the whorls of dicots while monocots have mostly three organs per whorl.

Tulipa gesneriana (tulip). A. top view of flower, B. and C. pistil overall view, B. cross section through the ovary; D. generalized flower diagram. Flower diagrams were drawn in the last century by the German botanist A. EICHLER (1875,

1878) for most of the indigenous and foreign plant groups. They symbolize the position, number and the symmetry ratios of the flower's single components in every circle of petals (according to W. TROLL, 1975).

The petals can often be traced back to reorganized stamens. This interpretation is proven by the fact that links between both categories of structures exist. A classic example is the white waterlily (Nymphaea alba, a rather primitive angiosperm) but such metamorphosis can, too, be found in many other species. Cultivators use this feature to select filled flowers. The wild rose (Rosa canina), for example, has only five petals and many stamens while most of the cultured varieties are characterized by filled flowers with many petals and a reduced number of fertile stamens.

A normal stamen consists of a pollen-containing, fertile anther (a group of two to four microsporangia) borne at the tip of a blade stalk or filament. The anther has (usually) two theces that are combined via the connective. Each theca includes two pollen sacs where the production of the pollen takes place.

Sterile stamens are called staminodes. One variation are the nectar-secreting honey leaves that live mostly between perianth and stamens. They can adopt the size and colour of petals and belong to the pollinator-attracting apparatus of the plant, like in buttercup (Ranunculus, Ranunculaceae).

The central part of a typical angiosperm flower is the gynoecium that is composed of one (simple pistil) or more (compound pistil) carpels. The carpels are, contrary to those of the gymnosperms, fused (see beneath). Their original structure that of a leaf, is hardly recognizable. The gynoecium contains one or more pistils.

A characteristic pistil has, from bottom to top, an enlarged basal ovary that includes the embryo sac, a columnar style and distal stigma, the organ that receives the pollen. Stigmata may be very varied in shape and structure. They are often button-shaped and equipped with papills. It is not rare that they are branched. It is distinguished between dry and moist stigmata. The surface of the dry ones is often studded with hair-shaped papills while moist ones are coated with a sticky film. The style serves to bring the stigma into a favourable position for pollination.

The leafy character of the carpels can easily be seen with a cross-section through the ovary. Both rims of a carpel seem to be rolled in and fused so that a tube-shaped structure is formed. On the basis of the thus built chamber (locule) are the seeds. This protection is missing in gymnosperms. Here the seeds lie openly on the carpels.

The suture that is formed by the fusion of the rims is called the ventral suture in angiosperm carpels while the midrib of the carpel is somewhat misleadingly called the dorsal suture.

Normally the gynoecium of angiosperms is composed of several carpels. A remarkable exception is made by the leguminoses where the flower contains just one carpel. Each carpel of a gynoecium may either form a pistil of its own (choricarpy or apocarpy) or several carpels may be fused together to one pistil (coenocarpy). The advantage of the latter solution is in the ability to distribute the pollen tubes among all carpels after successful pollination.

The way in which carpels are fused together and the resulting position of the ovules within the ovary, the so-called placentation, has been recognized to be an important taxonomical feature for the classification of angiosperms by A. ENGLER and E. PRANTL at the end of the last century. Some orders have been named after their moulding of ovary architecture, like for example centrospermae or parietales.

An ovary that is built from free carpels is called apocarp and is supposed to be the most original form of an angiosperm gynoecium. It is often found in ranunculaceae. Ovaries with fused carpels are derived from it.

In a syncarp ovary, the carpels are fused laterally. Its number of locules fits that of its carpels that are separated from each other by a septum. This can be seen very clearly with the ovary of a tulip. The ovules are inserted at the inner angles of the locules (axial placentation). False septums may have been put in later on which could lead to the misapprehension of a syncarpous ovary at superficial examination. But the unchanged parietal placentation of the ovules enables a definite classification.

Scheme of evolution for the basic type of gynoecium. The syncarpous gynoecium developed from the apocarpous one (on top) which again differentiated into the paracarpous (right) and lysicarpous (left) gynoecium. The paracarpous and the lysicarpous types are depicted in two different stages of evolution to elucidate the different ways of development from the basic syncarpous type. The ovules are given in dark brown (according to A. TAKHTAJAN, 1942).

Scanning Electron Micropgraph: cross-section through immature flower bud. Detail: Bilocular ovary with immature ovules, anther with immature pollen (tetrad state) - Snapdragon (Antirrhinum majus).

Lysicarpous ovaries display a unified cavity formed through disappearance of the lateral rims. Their ovules are inserted at an axial cone in the centre of the ovary (free central placenta).

The position of the ovary in relation to the other structures of a flower is an important taxonomical feature. It has to be distinguished between a hypogynous (the perianth is attached to the receptacle below the pistil), perigynous (perianth and stamens are borne on the rim of a concave structure in the depression of which the pistil is borne) or epigynous ovary (blossom seems to arise upon or above the ovary).

Flowers that contain both an androecium and a gynoecium are called androgynous or hermaphroditic. The flower is male, if just one androecium is present; if only one gynoecium is developed, then it is a female flower. A plant that has both male and female flowers is called monoecious while species where male and female flowers live on different plants, are called dioecious.

In the extreme, the flower is reduced to only one stamen or carpel. An especially interesting case is the spurge (Euphorbia) and some of its closer relatives. With them several extremely reduced flowers are grouped together to a functionally hermaphroditic unity, a cyathium. By integration of a set of bracts that bear great, often yellowish nectaries and some additional coloured bracts, the impression of a normal hermaphroditic flower is perfected (example: poinsettia; Poinsettia pulcherrima).

Inflorescences are arrangements of more than one flower, like bostryx, head, cyme etc. They are modified parts of the shoot with leaves typically reduced to leaflike bracts.

Only rarely does the main shoot or a side shoot end with a single, terminally placed flower. More common are flowers that are grouped together to units that are more or less clearly set off from the rest of the plant's body. These units are called inflorescences. An inflorescence is a flowering and thus modified part of the shoot with leaves that are usually reduced to leaflike bracts.

The types of branching occurring in flowering plants can be studied especially well with inflorescences. As has been said when reviewing the different types of branching of the shoot, it is distinguished between monopodial and sympodial inflorescences.

Some technical terms to start off: both determinate and indeterminate inflorescences occur. The first type of inflorescence ends with a flower at the top of the axis while this is missing with the latter. Indeterminate inflorescences can therefore theoretically continue to elongate and produce new flowers, while determinate inflorescences cannot. Indeterminate types of inflorescences are racemes, corymbs, heads, cones and umbels. They are also called simple inflorescences since their main axis is not branched. Complex types are branched. Among the complex inflorescences are double racemes (where a single flower is replaced by a raceme), double umbels, etc. as well as panicles and thyrsi.

Panicles are typically closed with lateral inflorescences that decrease continuously in branching towards the top of the inflorescence (partial inflorescence). With thyrsi, the single inflorescences are organized sympodially. Since in dicots only the first two leaf organs of a lateral axis can have the function of bractlets, a sympodially organized axis has a maximum of two lateral shoots. This generates a dichasium with an often fork-like appearance. If only one axial product of a bractlet is produced, it is spoken of a monochasium. Depending on whether the branching takes place in an alternating way (left and right) or only at one side, the resulting inflorescence is called a cincinnus (scorpioid cyme) or a bostryx (heliocoid cyme). The corresponding monocotyledon units where usually only one bractlet per lateral axis is developed are called rhipidium (fan-shaped cyme) and drepanium (helicoid cyme).

Fruits and Seeds

Fruits are a uniquely angiosperm feature: part of the pollinated flower ripens and becomes the fruit. Two types of fruits with a generally different mechanism of seed distribution exist: dehiscent and non-dehiscent fruits. Dehiscent fruits set the seeds free by opening of the fruit that remains itself with the mother plant. In contrast, non-dehiscent fruits and their seeds are a dispersal unit. Dehiscent fruit types are follicles, pulses, pods and capsules.

Non-dehiscent fruits are berries, drupes and nuts. Seeds are normally surrounded by a tough shell, the testa, that is derived from one or more integuments. It contains the endosperm, a nourishing tissue. Seeds are dispersed either by self-dispersal, also called autochory, or by allochory which means that extern factors are involved. Allochor modes of dispersal are dispersal by wind (anemochorys), water (hydrochory) or by animals (zoochory).

Fruits

Fructification is a feature of angiosperms during which the seeds are encapsulated by the ovary. Gymnosperms develop seeds that lie naked on the carpels.

A fruit is the state that the flower turns to during the ripening of the seeds. Perianth and stamens have usually withered and fallen off. During fructification, the gynoecium is largely rearranged, the ovary extends strongly, the stigma shrivels and the style normally degenerates or falls off but it, too, may develop further, if it serves to enlarge the fruit.

In the simplest case, the seeds are set free from the ovary after ripening by opening of the fruit (dehiscent fruits). But very often, both fruits and seeds form an integrated part of the dispersal unit (non-dehiscent fruit). If this happens, the tissue of the ovary is strongly differentiated. The pericarp is the part of a fruit that is formed by the ripening of the ovary wall. It is organized into three layers: the skin (exocarp), the often fleshy middle (mesocarp), and the membranous or stony inner layer (endocarp). These layers may become skin-like and leathery, fleshy or stringy (sclerotic) during fructification.

The fruits produce gadgets for the protection and the dispersal of the seeds. Additionally, other parts of the flower may contribute to the formation of the fruit, like the flower axis (strawberry) or the calyx. Depending on the organization of the gynoecium, single fruits or multiple fruits develop. Single fruits spring either from a monomer (coenocarp) gynoecium or from single carpels of an apocarp gynoecium while multiple fruits develop from flowers with a choricarp gynoecium.

If the fruit opens up to free ripe seeds, it is a dehiscent fruit where the pericarp is always dry. Contrasting are indehiscent fruits where the seed is spread together with the fruit or parts of the fruit.

The Different Dehiscent and Indehiscent Fruit Types

Follicle: It is regarded as the most primitive type of fruit and is common in the ranunculaceen family. The fruit is composed of a single carpel. Opening occurs usually at the ventral suture as with the fruits of larkspur (Delphinium). In rare cases, it does take place at the dorsal suture (Magnolia, for example).

Pulses: This is the type of fruit that occurs with leguminoses. Contrary to the follicles, both ventral and dorsal suture open at the same time. Both parts of the carpel wind normally up like a screw. The process may occur like an explosion so that the seed is hurled away with full force.

Pods are typical for cruciferous plants (Cruciferae, Brassicaceae). A pod stems from a paracarp ovary. It includes only two carpels and is divided by a central septum. When ripening, two flaps fall off while the septum and the placentas (replum) stay. The seeds may stick to them for a while.

Capsules are a very widespread type of fruit. They develop from a coenocarp gynoecium, the pericarp is dry. The opening of the fruit can take place in multiple ways by longitudinal tears, pores or by raising of the lid.

Indehiscent fruits do not scatter their seeds. It is distinguished between berries, drupes, and nuts depending on the composition of the pericarp.

The pericarp of berries becomes fleshy and juicy in all its parts and is often coloured. A berry like, for example, a tomato contains many seeds. The endocarp of drupes differentiates into a stone that encapsulates usually only one seed. The outer layers of the fruit (mesocarp and exocarp) are, just like with berries, fleshy. Well-known examples are cherries, plums, peaches, olives, and walnut.

Berries and drupes have adapted to endozoochory, *i.e.* they are eaten by animals (birds, mammals) along with the seeds. The seed (sometimes together with the stony endocarp) is excreted without being digested and is thereby scattered. The whole pericarp of nuts becomes hard and develops a thick wall which includes as a rule only one seed (hazelnut, acorn).

The caryopsis of sweet grass belongs to the nuts. Its seed is surrounded by a rather thin shell that is fused with the outer layer of the fruit. Other nut types are the achenes of composite flowers (Compositae, Asteraceae) where the calyx has been rearranged into a pappus (a structure for flying). An especially interesting case is the coconut (Cocos nucifera) where the outer layer is fibrous so that the fruit is able to swim due to the enclosed air.

Multiple fruits develop from flowers with an apocarp (choricarp) gynoecium. The single fruit may have a dry outer layer (single-seeded nuts or follicles with only few seeds) or the outer layer may be fleshy just like that of a drupe. Several fruits are usually combined to a unit by an axial structure that is either a central axial cone (strawberry, blackberry) or an axial cupule (rose hip, apple).

With some species, the numerous fruits that develop from the inflorescences stay together as a functional unit (syconium). Well-known examples are mulberry (Morus) and fig (Ficus).

Seeds

Seeds are normally surrounded by a tough shell (Testa) that develops from the integument (or several integuments) and that protects a largely developed embryo and its nutriments. The embryo is thus preserved from loss of water and bad weather. The nutriments are in a special nourishing tissue, the endosperm or in the perisperm (developed from tissue of the nucellus as with pepper). They have either to be made useable and to be used during germination or they are already stored in the cotyledons (beans, peas, almonds). Monocotyledons have normally only one cotyledon, dicots have two and gymnosperms several.

With many species, the seeds are unable to germinate immediately after falling off the plant (dormancy). This dormancy can be overcome by a period of cold weather (vernalisation) so that the seeds do not germinate before the beginning of the next period of vegetation.

The number of seeds that is produced by a single plant in the course of its life varies enormously. Some species like the coconut palm produce only few but very well provided descendants while others produce up to several million seeds that are inevitably provided rather badly. Numerous devices support the dispersal of the descendants since only a suitable habitat enables the seed to develop into a viable plant. If the dispersal is achieved only through devices that the plant itself produces, it is spoken of self-dispersal or autochory.

If extern factors such as wind, water, animals, etc. are involved, then the mode of dispersal is called allochor. Only rarely the seeds are spread on their own, usually the whole seed-containing fruit is scattered. Among the autochor modes of dispersal is flinging and spurting of seeds.

Dispersal by wind (anemochory) is widespread among plants. The units for dispersal are necessarily either very small and light and / or equipped with specially formed devices for flying or floating. Those devices are, for example, air-filled cavities in the seeds of orchids, seeds that are completely covered with hairs like cotton, pappus-bearing seeds (composite flowers) and units that are equipped with all different kinds of wings.

Dispersal by water (hydrochory) is much less common. Seeds that are spread by water have often an outer layer that cannot be moistened or they are hairy in order to trap air bubbles. Some big fruits that are spread by ocean currents like the coconut are surrounded by a loosely structured, airy layer.

Zoochory, *i.e.* the dispersal of fruits with the aid of animals has already been met when talking about berries, drupes, and multiple fruits. If seeds are eaten and excreted again, this is called endozoochoric dispersal. Contrasting is epizoochoric dispersal where the seeds or fruits stick to the animal or its fur either by slimy secretings or by variably shaped barbs.

A speciality in the dispersal through animals is that through ants (myrmecochory). Such seeds or fruits bear attachments, the elaiosomes that contain lures and nutriments. Myrmecochory is common with plants that live at the forest soil like violets (Viola).

The given examples show that zoochory offers a variety of possible mechanisms for the dispersal of seeds. But this variety is far beyond that of the architectural design of flowers. This is due to the fact that specialization and adaptation to seed or fruit dispersing animals is far less advanced than the specialisation of pollination.

Vascular Bundles of Roots

The vascular bundles of roots are enclosed by two concentric tubes of cells, the pericycle and the endodermis. The whole complex is embedded in the primary bark. Casparian strips are a typical feature of the vascular bundles of roots. Roots are able to perform secondary growth, which starts with the development of secondary cambium and leads to a restructuring of the vascular bundles. A feature of roots with secondary growth are feeder roots.

The vascular bundles of roots are constructed rather simply and show but little variations, if compared to those of the shoot. One reason could be the less variable surrounding that poses fewer demands than the surrounding above ground. Xylem- and phloem elements are combined in a central strand without pith, the stele. The root of dicots is often regarded as the basic type. The root of monocots differs only negligibly as far as the primary construction is concerned so that a separate reviewing is unnecessary.

The roots of many dicot species like wood plants or perennial plants are able to perform secondary growth, while roots of monocots can, with few exceptions like Dracaena, not grown in this way. The primary cylinder of vascular bundles is as a rule composed of radially located xylem elements and peripheral phloem. The vascular elements are enclosed by two concentric tubes of parenchyma tissue, the pericycle (inner cylinder) and the endodermis (outer cylinder). The pericycle is missing in the roots of some simple aquatic plants and in some parasitic species. The whole complex is embedded into a parenchyma of large cells the primary bark.

The radial walls of the endodermis are marked by characteristic thickenings, the Casparian strips that were discovered by ROBERT CASPARI, a German botanist, in 1865/66. These thickenings contain lignin, suberin and other encrusting substances, among them sometimes also phenolic oxidation products that lend the strips a dark colour. The endodermis is regarded as a typical cell layer of roots, but it is frequently found in shoots, too, where it, just as in roots, is localized at the periphery of the vascular cylinder. In the shoots of some ferns surrounds the endodermis single vascular bundles.

The relation of the vascular bundle arrangements of roots and shoots is illustrated by the theory of steles. It is based upon the assumption that root and shoot are merely two parts of one axial unity. A continuously recurring problem is the question of the vascular bundles' reorganization in the zone of transition between both organs. It should, strictly spoken, not be talked of a root to shoot transition, since the two organs are separated by both epicotyl and hypocotyl. These are formed first in the development of the embryo and only later are shoot and roots laid out. During this early phase, vascular bundles are formed in the area of the epi- or hypocotyl that elongate, beginning at the point of initiation, simultaneously both in the direction of the shoot and the roots and in this process adopt the organ-specific arrangement.

Secondary growth occurs, as has already been mentioned above, in the roots of many dicots. In the course of this, as a rule a secondary cambium develops out of parenchyma cells that have been laid out during primary growth between the xylem rays. The cambium is supplemented by cells of the pericycle so that finally a homogeneous cambium cylinder is formed. As a result of the secondary growth, pericycle and epidermis tear apart. The consequent gaps are filled by division of additional parenchyma cells. Just like shoots, roots develop a secondary final tissue (phelloderm, phellogen, phellem).

Although the vascular bundles of roots and shoots share many principles of organization, some fundamental differences exist. The formation of the shoot's vascular bundles is normally linked to the simultaneous lay out of leaves or branches and starts at the apical meristem. Lateral roots spring from cells of the pericycle and grow from the inside towards the outside. Only in a second, independent step of differentiation is the contact of the newly developed vascular bundles to the primary (tap) root established. The lateral roots may branch again. In this way, side roots of 2., 3.,.... order develop.

Contrary to leaves, side roots are built with central symmetry in a strictly morphological sense (exceptions are, for example, the roots of Kalanchoe), while leaves and leaf stalks are normally organized dorsiventrally. But a physiological dorsiventrality seems also to exist in lateral roots (see geotropism).

It is known that roots perform two in some respects contradictory tasks. On one hand they anchor the root in the soil, on the other hand they take up water and nutrients from the soil. The anchor function demands a tension-proof root, interspersed with strengthening elements like lignification or fibres within the phloem. This function is supported by secondary growth. On the other hand lignification and formation of bark decreases the ability for water uptake and transport.

Therefore has, also in roots, the 'division-of-labour-principle' been put through. Perennial plants in particular have at the endings of strongly lignified lateral or primary roots unlignified, short-lived branchings, called feeder roots. Additionally the secondary final tissues are not completely water-tight. How far they are engaged in the uptake of nutrients is not yet known.

Beside the primary and lateral roots, many mono- and dicots are equipped with so-called adventitious roots. They originate in different parts of the shoot or root. The root system of many Poaceae, for example, consists nearly entirely of adventitious roots (H. GUTTENBERG, Universitat Rostock, 1940). Often are roots or parts of them including the hypocotyl used as storage organs. The parenchyma tissue of the primary bark, the xylem and the phloem swell up and numerous cell divisions take place. The newly formed deposits are secondarily supplied with vascular bundles.

The Leaf

Although chloroplasts are found in the cells of young stems and immature fruits, leaves are the real photosynthetic factories of the plant.

A cross section through the blade of a typical leaf reveals 4 distinct tissue layers:

1. *Upper Epidermis:* This is a single layer of cells containing few or no chloroplasts. The cells are quite transparent and permit most of the light that strikes them to pass through to the underlying cells. The upper surface is covered with a waxy, waterproof cuticle, which serves to reduce water loss from the leaf.
2. *Palisade Layer:* This consists of one or more layers of cylindrical cells oriented with their long axis perpendicular to the plane of the leaf. The cells are filled with chloroplasts (usually several dozen of them) and carry on most of the photosynthesis in the leaf.

3. *Spongy Layer:* Lying beneath the palisade layer, its cells are irregular in shape and loosely packed. Although they contain a few chloroplasts, their main function seems to be the temporary storage of sugars and amino acids synthesized in the palisade layer. They also aid in the exchange of gases between the leaf and the environment. During the day, these cells give off oxygen and water vapour to the air spaces that surround them. They also pick up carbon dioxide from the air spaces. The air spaces are interconnected and eventually open to the outside through pores called stomata (sing., stoma).
4. *Lower Epidermis:* Typically most of the stomata (thousands per square centimeter) are located in the lower epidermis. Although most of the cells of the lower epidermis resemble those of the upper epidermis, each stoma is flanked by two sausage-shaped cells called guard cells. These differ from the other cells of the lower epidermis not only in their shape but also in having chloroplasts. The guard cells regulate the opening and closing of the stomata. Thus they control the exchange of gases between the leaf and the surrounding atmosphere.

Leaf Veins

Not only must the cells of the palisade and spongy layers be close to their air supply

- to secure CO_2 and
- release O_2
- (and the reverse in the dark) but they must be close to a leaf vein with its
- xylem to supply water and minerals
- phloem to remove synthesized food

The photo shows the network of leaf veins in a maple leaf. Probably no cell in the spongy layer is more than two cells away from a vein.

The xylem and phloem of veins is often surrounded by layers of sclerenchyma cells. These impart strength to the vein providing a stiff framework to support the soft tissues of the leaf blade.

Structure of Roots

The Root Tip: The root tip consists of a

- Meristem—a region of rapid mitosis, which produces the new cells for root growth.
- Root Cap—a sheath of cells that protects the meristem from abrasion and damage as the root tip grows through the soil.

Because of the frequency of mitosis in the meristem, root tips are often used to demonstrate mitosis in the laboratory The inset is a photo (courtesy of Carolina Biological Supply Co.) of anaphase in the meristem of an onion root tip.

The Region of Elongation: Here the cells produced by mitosis undergo a period of elongation in the direction of the axis of the root. It is at this time that they are sensitive to gravity and respond with gravitropism.

The Region of Differentiation: Here develop the differentiated tissues of the root:

- *Epidermis:* A single layer of flattened cells at the surface. When first formed, epidermal cells have extensions—the root hairs—which greatly increase the surface area available for the uptake of nutrients from the soil. The photo below shows the root hairs in the region of differentiation of a germinating radish seed.
- *Cortex:* A band of parenchyma cells that develops beneath the epidermis. It stores food. Its inner surface is bounded by a single layer of cells, the
- Endodermis.
- Stele
  - o Pericycle - the outer boundary of the stele. Secondary roots branch from it.
  - o Xylem—arranged in bundles in a spokelike fashion
  - o Phloem—alternates with xylem
- *Cambium :* In older parts of the root, another meristem forms between the xylem and phloem. Mitosis in the cambium produces new "secondary xylem" to the inside and secondary phloem to the outside.

Water Uptake: Water enters the root through the epidermis. Once within the epidermis, water passes through the cortex, mainly travelling between the cells. However, in order to enter the stele, it must pass through the cytoplasm of the cells of the endodermis.

Once within the stele, water is free again to move between cells as well as through them. In young roots, water enters directly into the xylem. In older roots, it may have to pass first through a band of phloem and cambium. It does so by travelling through horizontally-elongated cells, the xylem rays.

Mineral Uptake: One might have expected that minerals would enter the root dissolved in water. But, in fact, minerals enter separately:

- Even when no water is being absorbed, minerals enter freely - mostly through the root hairs.
- Minerals can enter against their concentration gradient; that is, by active transport. For example, plants can take up K^+ from the soil against a ten-thousand-fold concentration gradient; *e.g.*, from as little as 10 µM in the soil to 100 µM in the cell.
- Anything that interferes with the metabolism of root hairs interferes with mineral absorption.
- The root hairs are also the point of entry of mycorrhizal fungi. These transport minerals - especially phosphorus - to the root hair in exchange for carbohydrates from the plant.
- In legumes, the root hairs are the point of entry of rhizobia that will establish the mutualistic partnership enabling the plant to convert atmospheric nitrogen into protein.

Plants absorb their nutrients in inorganic form.

For examples:

- nitrogen enters as nitrate (NO_3^-) or ammonium ions (NH_4^+)
- phosphorus as PO_{43}^-
- potassium as K^+
- calcium as Ca_2^+

When you hear of the virtues of organic fertilizers, remember that such materials meet no nutritional need of the plant until their constituents have been degraded to inorganic forms. Organic matter does play an important role in making good soil texture, but only to the extent that it can yield inorganic ions can it meet the nutritional needs of the plant.

Once within the epidermis, inorganic ions pass inward from cell to cell, probably through plasmodesmata. The final step from the cytoplasm of the pericycle cells to the xylem is probably accomplished once again by active transport.

Gas Exchange: The older parts of roots are sheathed in layers of dead cork cells impregnated with a waxy, waterproof (and airproof) substance called suberin. This sheath reduces water loss but is as impervious to oxygen and carbon dioxide as it is to water.

However, the cork is perforated by nonsuberized pores called lenticels. These permit the exchange of oxygen and carbon dioxide between the air and the living cells beneath.

Fungi

Approximately 100,000 species of fungi have been identified.

Characteristics of Fungi

- Most fungi grow as tubular filaments called hyphae. An interwoven mass of hyphae is called a mycelium.
- The walls of hyphae are often strengthened with chitin, a polymer of N-acetylglucosamine.

The linkage between the sugars is like that of cellulose and peptidoglycan and produces the same sort of structural rigidity.

- Fungi are heterotrophic
    - o Some live as saprophytes, getting their nourishment from the surroundings (often having first digested it by secreting enzymes). They perform a crucial role in nature by decomposing dead organisms are releasing their nutrients for reuse by the living.
    - o Some live in a mutually beneficial symbiotic relationship with another organism, often a plant. The association of fungus and root is called a mycorrhiza.

- The plant benefits by more-efficient mineral (nitrogen and phosphorus) uptake.
- The fungus benefits by the sugars translocated to the root by the plant.

Mycorrhizal fungi may also form conduits for nutrients between plant species. The colourless, and hence heterotrophic Indian pipe (pictured on the right) is an angiosperm that must secure all its nourishment from mycorrhizal fungi that are attached at the same time to the roots of some autotrophic plant such as a pine tree. Radioactive carbon administered to the pine (as CO_2) soon turns up in the Indian pipes even though they may be growing some distance away on the forest floor.

- o Some fungi are parasitic, causing serious damage to their host (a few examples are given below).
    - Fungi disperse themselves by releasing spores, usually windblown. Fungal spores are present almost everywhere (and are a frequent cause of allergies). Spores of the wheat rust fungus have been found at 4000 m in the air and more than 1450 km (900 miles) from the place they were released. No wonder then that most fungi are worldwide in their distribution.

Ascomycetes (Phylum Ascomycota)

Ascomycetes produce two kinds of spores:

- asexual spores called conidia
- ascospores produced following sexual reproduction. Four or eight ascospores develop inside a saclike ascus (the group is commonly called sac fungi).

Some notable examples:

- Saccharomyces cerevisiae one of the budding yeasts. It ferments sugar to ethanol and carbon dioxide [Discussion] and thus is used.
    - o to make alcoholic beverages like beer and wine
    - o to make ethanol for industrial use
    - o in baking (it is often called baker's yeast). Here, it is the carbon dioxide that is wanted (to make bread and cakes "rise" and have a spongy texture).

Baker's yeast is also used:

- o in the commercial production of some vitamins.
- o in the production - using recombinant DNA technology- of some human therapeutic proteins.
  - Neurospora crassa, another favourite "model" organism in the laboratory.
  - The fungal partner in most lichens is an ascomycete.
  - Powdery mildews that attack ornamental plants
  - The chestnut blight, which in a few decades killed almost all of the mature American chestnut trees in the Appalachians of North America.
  - The Dutch elm disease, which has killed many of the American elms in the United States.
  - Pneumocystis jirovecii, which is a major cause of illness in immunosuppressed people, *e.g.*, patients with AIDS.
  - The truffle and the morel, both highly-prized food delicacies. The photo (courtesy of the French Embassy Press & Information Division) shows a farmer from the Perigord (in southwestern France) admiring a truffle. Truffles establish a symbiotic relationship with the roots of such trees as oaks.

Lichens

- Lichens are fungi that live in a symbiotic association with an autotrophic green alga or cyanobacterium (the "photobiont") or - in some cases - both.
- The fungal partner (the "mycobiont") in most lichens (98% of them) is an ascomycete. Zygomycetes make up the remainder.
- The relationship is often characterized as mutualistic; that is, both partners benefit. But recent evidence (*e.g.* in British soldier) suggests that while the fungus is dependent on its autotrophic partner, the photobiont is often perfectly content to live alone.

- Lichens secrete a variety of unusual chemicals; some of these probably assist in the breakdown of rock substrates like the one shown here.

The lower image is of the colourful lichen called British soldier.

- The fungus is Cladonia cristatella, an ascomycete. Its name is the name given to the lichen.
- The photobiont is Trebouxia erici, a green alga. It is found in many other lichens as well, and also can be found growing independently.
- The algal cells eventually are killed by the fungus, but are continuously replaced by new ones.
- So the relationship in this lichen is one of controlled parasitism rather than mutualism.
- The red cap produces the spores of the fungus, but these alone cannot form new lichens.
- Other structures (*e.g.*, soredia), containing both partners, are needed to disperse the lichen to new locations.
- Some lichens release only fungal spores. These mycobionts depend for their continued survival on finding an acceptable photobiont released from other lichens.

Phylogenetic trees, based on both ribosomal RNA genes and many protein-coding genes, as well as fossils indicate that lichens have been present on the earth for at least 600 million years.

Today about 14,000 species of fungi form lichens.

Lichens are extremely sensitive to air pollution. One of the best contemporary examples of evolutionary adaptation is the change in coloration of the peppered moth as the lichens in its habitat declined because of air pollution and then returned when air quality controls were put in place.

Some modern fungi (*e.g.*, Penicillium chrysogenum, the source of the antibiotic penicillin) appear to have evolved from lichen-forming ancestors - abandoning their original symbiotic way of life.

Basidiomycetes (Phylum Basidiomycota)

Basidiomycetes include mushrooms, shelf fungi, puffballs, rusts, and smuts. They are dispersed by spores borne at the tips of basidia (giving rise to the name for the group).

Mushrooms are masses of interwoven hyphae growing up from the main mass of the mycelium growing underground. The basidia develop on the undersides and release their spores (four from each basidium) into the air.

A single mycelium may expand outward year after year as its hyphae grow into new terrain. In some species, mushrooms are sent up once a year at the periphery producing a circle known since medieval times as a "fairy ring".

Some notable basidiomycetes:

- *Armillaria Bulbosa:* A single specimen in northern Michigan (USA) was found to have spread over 37 acres of the forest floor. RFLP analysis of samples taken from many different locations within this area showed that all the samples were from a single clone. Assuming the normal rate of vegetative growth for this species, it must have taken 1500 years to spread to that size.
- the cultivated, edible mushroom that finds its way into pizza, soups, etc.
- Amanita muscaria. Forms a beautiful mushroom but deadly when eaten.
- Smuts. Parasites of important crops like wheat, oats, and rye.
- Rusts. Some, such as
  - o wheat black stem rust (Puccinia graminis) and
  - o white pine blister rust are serious pests. Both have complicated life cycles during which they pass through a second plant host (barberry plants for wheat black stem rust, gooseberries or wild currants for the white pine blister rust).

Zygomycetes (Phylum Zygomycota)

All the fungi assigned to this group (which probably does not represent a single clade) form spores in a sporangium.

Some notable examples:

- The bread mold, Rhizopus stolonifera
- Rhizopus oryzae, used to make sake, the rice wine of Asia. Can also infect humans, especially if they are immunosuppressed (*e.g.*, AIDS patients, transplant recipients).
- Another species of Rhizopus is used in the commercial production of glucocorticoids.
- Many mycorrhizal fungi belong to this group.

6

Garden

A garden is a planned space, usually outdoors, set aside for the display, cultivation, and enjoyment of plants and other forms of nature. The garden can incorporate both natural and man-made materials. The most common form is known as a residential garden. Western gardens are almost universally based around plants.

Zoos, which display wild animals in simulated natural habitats, were formerly called zoological gardens. Some traditional types of eastern gardens, such as Zen gardens, use plants sparsely or not at all. Food-producing gardens are distinguished from farms by their smaller scale, more labour-intensive methods, and their purpose (enjoyment of a hobby rather than produce for sale). The gardening article discusses the differences and similarities between gardens and farms in greater detail.

Gardening is the activity of growing and maintaining the garden. This work is done by an amateur or professional gardener. A gardener might also work in a non-garden setting, such as a park, a roadside embankment, or other public space. Landscape architecture is a related professional activity with landscape architects tending to specialise in design for public and corporate clients.

Garden Planning and Design

Garden planning and garden design may be undertaken by a professional. A landscape architect is a professional who can plan and realise outdoor spaces. A garden designer is usually trained to plan and realise residential gardens.

The planner must give consideration to many factors:

- Purpose
- Existing conditions
- Financial constraints
- Maintenance implications

Elements of a Garden

The elements of a garden consist of natural conditions and materials, as well as man-made elements:

Natural conditions and materials:

- Soil
- Rocks
- Light conditions
- Wind
- Precipitation
- Air quality
  - o Pollution
  - o Proximity to ocean (salinity)
- Plant materials

Man-made elements:

- Terrace, patio, deck
- Paths
- Lighting
- Raised beds
- Outdoor art/sculpture, such as Gazebos
- Pool, water garden, or other water elements

Uses for the Garden Space

A garden can have many purposes- aesthetic, functional, and recreational. Uses for the garden space are:

- Cooperation with nature
  - o Plant cultivation
- Observance of nature
  - o Bird- and insect-watching
  - o Reflection on the changing seasons

- Relaxation
    - o Family dinners on the terrace
    - o Children playing in the yard
    - o Reading and relaxing in the hammock
    - o Maintaining the flowerbeds
    - o Pottering in the shed
    - o Basking in warm sunshine
    - o Escaping oppressive sunlight and heat
- Growing useful produce
    - o Flowers to cut and bring inside for indoor beauty
    - o Fresh herbs and vegetables for cooking

Types of Gardens

Fernery: A fernery is a specialized garden for the cultivation and display of ferns.

In many countries ferneries are indoors or at least sheltered or kept in a shadehouse to provide a moist environment and filtered light.

Flower Garden: A flower garden is a form of garden usually grown for decorative purposes, centering primarily on the kinds of flowers produced by the plants involved. Because flowers bloom at varying times of the year, and some plants are annual, dying each winter, the design of flower gardens can be sophisticated, taking such matters into consideration to keep blooms, even of specific colour combinations, consistent or present through varying seasons.

Flower gardens combine plants of different heights, colours, textures, and fragances to create interest and delight the senses. Photo by Ashley Sheets, provided courtesy of Park Seed Company.

These have grown in complexity over the years, and are sometimes tied in function to other kinds of gardens, like knot gardens or herb gardens, many herbs also having decorative function, and some decorative flowers being edible.

One simpler solution to flower garden design, growing in popularity, is the pre-planned Wild-flower seed mix.

Assortments of seeds are created which will create a bed that contains flowers of various blooming seasons, so that some portion of them should always be in bloom. The best mixtures even include combinations of perennials and biennials, which may not bloom until the following year, and also annuals which are "self-seeding", so they will return, creating a permanent flowerbed.

History: Many, if not most, plants considered decorative flowers originated as weeds, which if attractive enough would sometimes be tolerated by farmers because of their appeal. This led to an artificial selection process, producing ever-prettier (to humans) flowers.

Once domesticated, though, most flowers were grown either separately or as part of gardens having some other primary function. In the West, the idea of gardens dedicated to flowers did not become common until the 19th century, though in fact many modern gardens are indeed flower gardens.

Flower gardens are, indeed, a key factor in modern landscape design and even architecture, especially for large businesses, some of which pay to have large flower gardens torn out and replaced entirely each season, in order to keep the colour patterns consistent.

A herb garden is a garden specifically designed and used for the cultivation of cooking and/or medicinal herbs.

Herb gardens developed from the general gardens of the ancient classical worlds, used for growing vegetables, flowers, fruits and medicines. During the medieval period, monks and nuns acquired specialist medical knowledge and grew the necessary herbs in specialist gardens. Typical plants were rosemary, parsley, sage, marjoram, thyme, mint, rue, angelica, bay and basil. With the advance of medical and botanical sciences in Renaissance Europe, monastic herb gardens developed into botanical gardens. The section in which herbs was grown became known as a Garden of Simples.

Herb gardens experienced a revival with the work of the British garden historian and horticultural, writer Eleanour Sinclair Rohde (1882-1950). Modern herb gardens may be purely functional or may be ornamental, sometimes as part of a design and containing boxes and raised beds. The development of alternative medicine is also encouraging people to grow and use fresh herbs (*e.g.*, for the treatment of acne).

Lawn

A lawn is an area of recreational or amenity land planted with grass, and sometimes clover and other plants, which are maintained at a low, even height.

Usage: Lawns are a standard feature of ornamental private and public gardens and landscapes in much of the world today. Lawns are created for aesthetic use in gardens, and for recreational use, including sports. They are typically planted near homes, often as part of gardens, and are also used in other ornamental landscapes and gardens. Lawns are frequently a feature of public parks and other spaces. They form the playing surface for many outdoor sports, reducing erosion and dust as well as providing a cushion for players in sports such as football, cricket, baseball, golf, tennis, bocce and stake. In sports venues, the word lawn is often replaced by turf or green.

Many different species of grass are used, often depending on the intended use of the lawn, with vigorous, coarse grasses used where active sports are played, and much finer, softer grasses on ornamental lawns, and partly on climate, with different grasses adapted to oceanic climates with cool summers, and tropical and continental climates with hot summers. It is also not uncommon to mix grass seeds. A 50/50 mixture of grass types can, for example, form a stronger lawn when one grass type does better in the warmer seasons and the other is more resistant to colder weather.

Types of Lawngrass

There are thousands of varieties of lawngrass, each adapted to specific conditions of precipitation, temperature, and sun/shade tolerance. Breeders are constantly creating new and improved varieties of the base list of lawngrass species. The two basic categories are cool season grasses and warm season grasses.

Cool season grasses start growth at 5 °C, and grows at their fastest rate when temperatures are between 10-25 °C (Huxley 1992), in climates that have relatively mild/cool summers, with two periods of rapid growth in the spring and fall. They retain their colour well in extreme cold and typicaly grow very dense, carpetlike lawns with relatively little thatch.

- bluegrass
- bentgrass
- ryegrasses
- fescues
- Grama grass
- Princess 77 Bermuda (sometimes used in stadiums)

Warm season grasses only start growth at temperatures above 10 °C, and grow fastest when temperatures are between 25 °C and 35 °C, with one long growth period over the spring and summer (Huxley 1992).

They often go dormant in cooler months, turning shades of tan or brown. Many warm season grasses are quite drought tolerant, and can handle very high summer temperatures, although temperatures below -15 °C can kill most warm season grasses.

- zoysiagrass
- bermudagrass
- St. Augustine Grass
- bahiagrass
- centipedegrass
- carpetgrass
- buffalo grass

United States

Lawns cover a significant area in the United States. In a recent NASA-sponsored study, researcher Christina Milesi estimated the area covered by lawns to be about 128,000 square kilometers (nearly 32 million acres) making it the largest irrigated crop in the United States, about three times that of irrigated corn.

Lawn care is thus a major business in America. Maintenance, construction and management of lawns of various kinds are the focus of much of the modern horticulture industry. Estimates of the amount spent on professional lawn care services vary, but a Harris Survey put the total at $28.9 billion in 2002, which averages to around $1,200 per household, spread over the 24.7 million households who use such services.

Virginia Scott Jenkins, in her book The Lawn: A History of an American Obsession (1994), traces the desire to kill weeds historically. She notes that the current rage for a chemically-dependent lawn emerged after World War II, and argues that "American front lawns are a symbol of man's control of, or superiority over, his environment."

Approximately 50-70 percent of U.S. residential water is used for landscaping, most of it to water lawns.

Along with trees, lawns are a vital element in the fight against urban heat islanding. Lawns provide:

- Oxygen conversion,
- Filtering of air particulates,
- Erosion control,
- Air and surface cooling to offset asphalt, cement, and rooftops.

In comparison to bare dirt, a lawn may be 20 degrees cooler on a hot day, and up to 40 degrees cooler than cement surfaces.

History: Before the invention of mowing machines in 1830, lawns were managed differently from today. Lawns belonging to wealthy people were sometimes maintained by the labour-intensive methods of scything and shearing. In most cases however, they were pasture land, maintained by grazing with sheep or other livestock. Areas of grass grazed regularly by rabbits, horses or sheep over a long period can form a very low, tight sward which is similar to a modern lawn. This was the original meaning of the word "lawn", and the term can still be found in place-names. Some forest areas where extensive grazing is practiced still have these semi-natural lawns. For example, in the New Forest, England, such grazed areas still occur commonly and are still called lawns, for example Balmer Lawn.

Lawns became popular in Europe from the Middle Ages onward. The early lawns were not always distinguishable from pasture fields. It is thought that the associations with pasture and the biblical connotations of this word made them attractive culturally.

By contrast, they are little known or used in this form in other traditions of gardening. In addition, the damp climate of maritime Western Europe made them easier to grow and manage than in other lands. Lawns were also used in medieval times within monasteries and in the courtyards of castles for the lords and ladies to take their daily constitutional and escape from the odours of the castle.

It was not until the Tudor and Elizabethan times that the garden and the lawn became a place to be loved and admired. Created as walkways and for play areas, the lawns were not as we envisage them today. They were made up of meadow plants, such as camomile, a particular favourite.

In the early 1600s, the Jacobean epoch of gardening began. It was during this period that the closely cut "English" lawn was born. By the end of this period, the English lawn was the envy of even the French. It was also seen as a symbol of status by the gentry. In the early 1700s, gardening fashion went through a further change. William Kent and the age of Capability Brown were in progress, and the open "English" style of parkland was seen across Britain and Ireland. Lawns seemed to flow from the garden into the outer landscape.

During Victorian times, as more plants were introduced into Britain, and the influence of France and Italy became prevalent, lawns became smaller as borders were created and filled with plants, statues, sculptures, terraces and water features, which started eating into the area covered by the lawn. In the United States, it was not until after the Civil War that lawns began to appear in middle class residences. Most people did not have the hired labour needed to cut a field of grass with scythes.

Average home owners either raised vegetables in their yards or left them alone. If weeds sprouted that was fine. Toward the end of the 19th century, suburbs appeared on the

American scene, along with the sprinkler, greatly improved lawn mowers, new ideas about landscaping and a shorter work week.

Lawns do not have to be, and have not always been of grass. Other possible plants for fine lawns in the right conditions, are camomile and thyme. Some lawns, if grown in difficult conditions for grasses, become dominated by whatever weeds can survive there; these include clovers in dry conditions, and moss in damp shady conditions.

Maintenance

Maintaining a rough lawn requires only occasional cutting with a suitable machine, or grazing by animals.

Higher quality lawns however require a number of operations. These may include:

- Mowing, to cut the grass regularly to an even height
- Scarifying and raking, to remove dead grass and prevent tufting
- Rolling, to encourage tillering (branching of grass plants), and to level the ground
- Top dressing the lawn with sand, soil or other material
- Spiking or aeration, to relieve compaction of the soil
- Watering, to prevent from going dormant and turning brown
- Fertilizing
- Pesticide application to manage weeds and insects

Seasonal Lawn Maintenance: Seasonal lawn care will vary to some extent depending on the climate zone and type of grass that is grown, whether cool season or warm season varieties. In general, however, there are recognized steps in lawn care that should be observed in any of these areas.

Spring or early summer is the time to seed, sod, or sprig a yard, when the ground is warmer. For a new lawn, adding a fresh load of topsoil to the ground is beneficial. Seeding the lawn is the least expensive way to plant, but it takes longer for the lawn to grow and usually needs daily watering, or the freshly-sprouted grass will die. Sodding is more expensive, but it will provide an almost instant lawn that can be planted in most climate zones in any season. Hydroseeding is a relatively quick and inexpensive method of planting. A nitrogen-based, slow-release fertilizer should be applied, as well as a weed killer and a product for lawn pest removal.

Summer lawn care requires raising the lawn mower for cool season grass, and lowering it for warm season lawns. Lawns will require longer and more frequent watering, best done in early morning to encourage a stronger root system. This is also the time to apply an all-purpose fertilizer. During the hot summer months, lawns may be susceptible to fungus disease. It's advisable to take a sod sample to a local landscape expert for testing and treating the yard, if necessary.

In the fall, lawns can be mowed at a lower height and thatch buildup that occurs in warm season grasses should be removed, although lawn experts are divided in their opinions on this. This is also a good time to add a sandy loam and apply fertilizer, one that contains some type of wetting agent. Cool season lawns can be planted in autumn if there is adequate rainfall. Lawn care in the winter is minimal, requiring only light feedings of organic material, such as green-waste compost, and minerals to encourage earthworms and beneficial microbes.

Criticisms: A number of criticisms of lawns are based on environmental grounds:

- Many lawns are composed of a single species of plant, or of very few species, which reduces biodiversity, especially if the lawn covers a large area. In addition, they may be composed primarily of plants not local to the area, which can further decrease local biodiversity.
- Lawns are sometimes cared for by using pesticides and other chemicals, which can be harmful to the environment if misused.
- Maintaining a green lawn often requires large amounts of water. This was not a problem in temperate England where the concept of the lawn originated, as natural rainfall was sufficient to maintain a lawn's health.

However the exporting of the lawn ideal to more arid regions of the world, such as the U.S. Southwest, has crimped already scarce water resources in such areas, requiring larger, more environmentally invasive water supply systems. Grass typically goes dormant during cold, winter months, and turns brown during hot, dry summer months, thereby reducing its demand for water. Many property owners consider this "dead" appearance unacceptable and therefore increase watering during the summer months. Grass can also recover quite well from a drought.

- In the United States lawn heights are generally maintained by gasoline-powered lawnmowers, which contribute to urban smog during the summer months. The U.S. Environmental Protection Agency found that in some urban areas, up to 5% of smog was due to pre-1997 small gasoline engines such as are typically used on lawnmowers. Since 1997, the EPA has mandated emissions controls on newer engines in an effort to reduce smog.
- Lawns use up vast areas of arable land that might otherwise be used for growing crops.

However, using ecological techniques, the impact of lawns can sometimes be reduced. Such methods include the use of local grasses, proper mowing techniques, leaving grass clippings in place, integrated pest management, organic fertilizers, and introducing a variety of plants to the lawn.

In addition to the environmental criticisms, some gardeners question the aesthetic value of lawns. One positive benefit of a healthy lawn is that of a filter for contaminants and to prevent run-off and erosion of bare dirt. Highway construction projects in the United States now routinely include replanting grasses on disturbed soils for this purpose, although they are not maintained as lawns.

Orangery

An Orangery was a feature of royal and aristocratic residences from the 17th to the 19th century. A type of greenhouse, with citrus trees being grown in tubs and wintering under cover, it originated from the Renaissance gardens of Italy, when glass-making technology enabled sufficient expanses of clear glass to be produced.

The Orangerie at the Palace of the Louvre, 1617, inspired imitations that were not eclipsed until the development of the modern greenhouse in the 1840s, which was quickly overshadowed by the architecture in glass of Joseph Paxton. Notable for his design of the Crystal Palace, his "great conservatory" at Chatsworth House was an orangery and glass house of monumental proportions.

The orangery, however, was not just a greenhouse but a symbol of prestige and wealth and a feature of the garden, in the same way as a summerhouse, folly or "Grecian temple". Owners would conduct their guests there on tours of the garden to admire not only the fruits within but the architecture without. Often the orangery would contain fountains, grottos, and an area in which to sit and eat.

Continental European Orangeries

- Chateau de Versailles
  - Versailles Orangerie
- Tuileries: Orangerie in the Tuileries Gardens, Paris
- Schonbrunn, Vienna
- Peterhof, Bolshaya Kamennya Oranzhereya (1722-25)
- Tsarskoe Selo, Bolshaya Oranzhereya (1762, 1820)
- Kuskovo, Moscow, Oranzhereya (illustration, right)
- Potsdam, Orangerieschloss (illustration, above right)
- Fulda, Orangerie
- Kassel, Orangerie
- Gera, Orangery and "Kuchengarten"
- Strasbourg, park of the Orangerie
- Hanover, a part of the Herrenhausen Gardens
- Nynas Slott, Manorial Estate (Castle) and Orangery, Nynas, Sweden

Orangeries in the UK

The orangery at Kew (1761) is the earliest surviving work there by Sir William Chambers. At 28 m (92 ft) long, it was the largest glasshouse in Britain when it was built. Though it was designed as an arcade with end pavilions to winter oranges, the light levels under its solid roof were too low for it to be successful.

- *Royal Botanical Gardens, Kew—The Orangery:* The Orangery at Margam Park, Wales, was built between 1787 and 1793 to house a large collection of orange, lemon and citron trees inherited by Thomas Mansel Talbot. The original house has been razed, but the surviving orangery, at 327 feet, is the longest one in Wales.
- *Margam Park—Orangery :* Sir Christopher Wren designed the orangery at Kensington Palace for Queen Anne in 1702.

There is an orangery dating from about 1700 at Kenwood House in London.

Orangeries in the US

In the United States the earliest surviving orangery is at the Tayloe house in Mount Airy, Virginia, but today it is an overgrown ruin. A ruined orangery can also be seen at the gardens Eyre Hall in North Hampton County, Virginia.

A more interesting, and extant, early 18th century orangery can be seen at the Wye Plantation, near Tunis Mills (Easton), Maryland. This orangery sits behind the main house and consists of a large open room with two smaller wings added at some point after the initial construction.

The south-facing wall consists of large triple-hung windows. A second story was added as a billiards room, according to the current resident, Ms. Tilghman, a descendent of the Lloyd family. This plantation is also notable as having been the home of [Frederick Douglas] as a young slave boy. The orangery is described in the book Glass Houses, as is the orangery at the Tayloe house.

Ms. Tilghman notes that plants are still stored inside the building in winter, but a frame has been constructed to hold the houseplants, and the whole of the frame is covered with plastic to keep in moisture. In this way, the plants do not have to be watered through the entire winter. A new orangery, loosely modeled on the Wye orangery, was built on private property in 2006 near Royal Oak, Maryland. This orangery is approximately 29' x 17' of frame construction on stab, with a 20 foot ceiling, and glass on both front and back.

Orchard

An orchard is an intentional planting of trees or shrubs maintained for food production. Most orchards comprise either fruit or nut-producing trees (see fruit trees), for commercial production. Orchards are also sometimes a feature of large gardens, where they serve an aesthetic as well as a productive purpose. Most temperate-zone orchards are laid out in a regular grid, with a grazed or mown grass or bare soil base that makes maintenance and fruit gathering easy.

Orchards are often concentrated near bodies of water, where climatic extremes are moderated and blossom time is retarded until frost danger is past.

The forest garden is a food production system that is closely related to the orchard. A move towards more ecologically-friendly coffee production has led to forest-garden production of coffee. Brazil Nuts and Rubber are being produced in such a method in some areas. Often, mixed orchards are planted. In Europe Quince is sometimes planted along with apples.

Crops

Tropical Areas

- Banana
- Coconut
- Cacao
- Coffee
- Guava
- Mango

- Papaya
- Tea

Subtropical Areas

- Date Palm
- Citrus (Grapefruit, Lemon, Mandarin, Orange, etc.)
- Lychee

Temperate Areas

- Apple
- Cherry
- Hazel
- Peach
- Pear
- Pecan
- Walnut
- Plum
- Apricots
- Persimmon

Orchards by Region

The most extensive orchards in the United States are apple and orange orchards, although citrus orchards are more commonly called groves. The most extensive apple orchard area is in eastern Washington state, while there are extensive orange orchards in Florida and southern California. A particular advantage of growing apples on the high plateau areas of Washington state is that it is possible to grow high-quality organic apples. In eastern North America many orchards are along the shores of Lake Michigan, Lake Erie, and Lake Ontario.

Murcia is a major orchard area in Europe, with citrus crops. New Zealand, China, Argentina, and Chile also have extensive apple orchards.

Rose Garden

Rose Garden may refer to:

- Any of a number of gardens which feature roses as the main plant life

- The International Rose Test Garden in Portland, Oregon
- The Rose Garden Arena in Portland, Oregon
- The Rose Garden (San Jose) neighbourhood in San Jose, California
- The White House Rose Garden
- "Rose Garden Project", the common name for the Port and Airport Development Strategy carried out in Hong Kong during 1990s, which includes the construction of Hong Kong International Airport
- Rosegarden, the Linux MIDI sequencer
- "(I Never Promised You a) Rose Garden", a 1970 Grammy-winning song for singer Lynn Anderson
- Rose Garden (Lynn Anderson album), an album by Lynn Anderson
- I Never Promised You a Rose Garden, a novel, and a 1977 film

Vegetable Garden

A vegetable garden (also known as a vegetable patch or vegetable plot) is a garden that exists to grow vegetables and other plants useful for human consumption, in contrast to a flower garden that exists for aesthetic purposes. It is a small-scale form of vegetable growing. A vegetable garden typically includes a compost heap and several plots or divided areas of land, intended to grow one or two types of plant in each plot. Many families have home kitchen and vegetable gardens that they use to make food. In world war 2, people had gardens called 'victory gardens' which provided food for the war effort.

White Garden: A White Garden is a feature garden comprised of plants that produce white flowers and spathes as well as plants with a white or silvery cast to their foliage. The white garden is a variant of the colour garden. The most essential feature of the white garden is its unity of colour.

The white garden is an informal gardening style that is similar in design to the English cottage garden. The open and informal design creates associations with romance, peace, and elegance. The white flowers are not usually placed in clusters, but spread throughout the garden's green areas, creating a natural look and feel. The mildly dense placement of white flowers creates a luminescent sight that is especially powerful in the twilight.

Winter Garden: A winter garden is, as the name implies a garden planted to either produce food, or at least remain visibly planted and slowly develop, throughout the winter, or else a garden whose plants will serve as living decoration all winter.

Vegetables which are typically, or can be, used in a winter garden include:

- Several breeds of winter hardy cabbage
- Certain winter hardy breeds of broccoli
- Winter rye is even grown where a summer garden will be, in order to protect the ground from weeds, and provide soil amendment when tilled directly into the soil the following spring
- Rosemary, which is evergreen and can even be grown as a christmas tree-shaped bush
- beets
- carrots
- alliums — onions, chives, and their relatives are evergreen, though some may die back during the winter and recover in the spring.

One basic premise, in temperate or colder regions, to the winter garden is that the plants may indeed become dormant when snow covers the ground, but will grow each time the sun heats at least part of the plant to above freezing (snow or not), especially in regions where snow cover and below-freezing temps are not constant for months at a time. Perhaps the most famous "winter garden" is at the Biltmore Estate near Ashville, North Carolina.

Natural Landscaping

Natural landscaping, also called native gardening, is the use of plants, including trees, shrubs, groundcover, grasses and wildflowers, which are indigenous to the geographical area in which the garden is located, as well as rocks and boulders in place of groomed lawns and planned planting beds to blend residential or commercial property into the natural surroundings of the particular area.

Benefits

Maintenance: In theory, natural landscaping is adapted to the climate, geography and hydrology and should require no pesticides, fertilizers and watering to maintain, given that native plants have adapted and evolved to local conditions over thousands of years. However, these applications may be necessary for some preventative care of trees and other vegetation.

Native plants suit today's interest in "low-maintenance" gardening and landscaping, with many species vigorous and hardy and able to survive winter cold and summer heat. Once established, they can flourish without irrigation or fertilization, and are resistant to most pests and diseases.

Many municipalities have quickly recognized the benefits of natural landscaping due to municipal budget constraints and reductions and the general public is now benefiting from the implementation of natural landscaping techniques to save water and create more personal time.

Bush regeneration shares many similarities, though it targets preexisting patches of (often heavily degraded) original bushland and has removal of weeds as a high (sometimes higher) priority than replanting of native plants.

Native plants provide suitable habitat for native species of butterflies, birds, and other wildlife. They provide more variety in gardens by offering myriad alternatives to the over-planted cultivars and aliens. These plants have co-evolved with animals, fungi and microbes, to form a complex network of relationships. They are the foundation of their native ecosystems, or natural communities.

Ecology

Native plants provide suitable habitat for native species of butterflies, birds, and other wildlife. They provide more variety in gardens by offering myriad alternatives to the over-planted cultivars and aliens. These plants have co-evolved with animals, fungi and microbes, to form a complex network of relationships. They are the foundation of their native ecosystems, or natural communities.

Many weeds in an area are usually the result of imported plants. These plants become invasive because there are no natural controls such as disease, weather, or fauna in their new environment. They take over native habitats, reducing shelter and food for local fauna. Using local provenance plants increases the biodiversity of and is important for the health of a region's overall ecology. Much of our wild areas have been destroyed to make room for development. Housing developments have replaced native habitats with houses and vast lawns. While development won't be stopped, gardeners can keep wild areas and green spaces filled with native species on their lots and in their communities.

Such gardens often benefit from the plants being habituated to the climate and soil conditions present and may require fewer soil amendments or chemical treatments in order to present a pleasing display.

Challenges: However, while local provenance plants have adapted to local conditions (which includes climate, soil, and other native plants and animals), there will often be instances, especially in cities, where one or more of these will have been radically altered.

Examples include:

- Building rubble used as landfill in a garden may result in a markedly more alkaline soil, which can be a problem if this is a region where the soil is usually acid (and hence the local plants are adapted to acid soils).
- Buildings often cast a substantial amount of shade, especially with garden spaces shrinking in most urban areas. This may give rise to conditions substantially shadier than many local plants would like.
- Soil which is high in organic material and nutrients is often introduced into gardens, or many gardeners will

have used fertilizers. Plants from some areas, *e.g.* many Australian plants are particularly sensitive to phosphorus.

Despite this, there are usually plenty of indigenous plants which will grow and thrive in the area one is trying to establish a native garden.

Types

- fully forested with leaf debris forest floor
- desert with arid loving plants and succulents
- grassy meadow with a variety of wildflowers and water features
- lowlands savannah with grasses and native trees
- ocean dunescape with tall, salt tolerant grasses

Wild-flower Gardens

Wild-flower is a term used in some countries to describe the numerous showy flowers from some drier climates, most notably southwest Western Australia, Southern Africa and North America.

Rain Gardens: Rain gardens that absorb rainwater from gutters and impervious surfaces, work much better when planted with native plants tolerant to the alternate flooding and drying cycles. Some Wild-flower gardens attempt to recreate a prairie landscape, including native grasses along with the flowering plants, or forbs. Such gardens benefit the local wildlife, often attracting birds, butterflies and small mammals. By carefully choosing the plants in the garden, the gardener can encourage some of these visitors to the garden. One popular type of Wild-flower garden specializes in attracting butterflies and is called a butterfly garden.

The native plants used in Wild-flower gardens often have deep root systems. This makes them good plants for absorbing runoff and allowing the water to filter back into the local water table. Wild-flower gardens that focus on capturing runoff in this fashion are called rain gardens.

Advantages

- no fertilization required
- no additional water

- more water available for other uses and other people
- zero to near zero work needed for maintenance
- no lawn mowing
- erosion reduced to a minimum
- natural landscaped plants take full advantage of rainfall
- when water restrictions are implemented, natural landscaped plants will survive, while more traditional plants may not
- increased habitat for native flora and fauna
- where heavily forested, provides shade on homes and businesses saving energy
- native plants rarely become invasive

Disadvantages

- not good for people and outdoor games that require a manicured turf.
- poisonous plants such as poison ivy and/or poison oak may make the habitat home.
- increased wild animal intrusion
- in certain areas, wildfires or bushfires may be of great concern.

New Construction

In new construction, builders can either avoid clear cutting or clearing an entire property and disturbing other large flora or builders can completely clear an area of all flora to save construction time and replace the clearing with juvenile specimens once the job is complete. The downside to this is additional costs involved with purchasing replacements. The builder may also choose to plant additional native trees and other flora after construction to help the property blend with natural surroundings. In some planned developments, natural landscaping is the requirement. Builders may not remove trees larger than a specific diameter and owners may not arbitrarily cut trees without a permit.

Reclaiming Land: Throughout the world, forested areas were turned into cattle grazing or farmland. Often this land is then turned into residential or commercial use property. By returning the land back to its original state prior to human disturbance, vast amounts of energy usage and increasing pollution can be reduced. Natural landscaping costs less to install than traditional landscaping and, after the initial few years, reduces maintenance costs, combats erosion, and accommodates storm and flood waters better.

Native Plant Societies: In many parts of the world, there are societies, clubs or local groups, such as Bushcare or ASGAP in Australia, which are made up of gardeners interested in growing plants local to their area, state or country. They can be the best way to find out about (and to obtain, by buying or swapping) local native plants. Many members have spent years or decades cultivating local plants or bushwalking in local areas.

Additional Sources

- Wildflowers and Weeds, by Booth Courtenay and James Zimmerman. paperback
- Plants of the Chicago Region, by Floyd Swink and Gerould Wilhelm.
- Field Guide to Wildflowers of Northeastern and North-central North America, by Roger Peterson.
- Audubon Society Field Guide to North American Wildflowers; Eastern Region, by William A. Niering. Alfred A. Knopf. Soft Cover.
- Newcomb's Wild-flower Guide, by Lawrence Newcomb. Little. Paperback.
- Field Guide to Trees and Shrubs, by George A. Petrides. Houghton., paperback.
- Field Guide to the Ferns, by Roger Tory Peterson. Paperback.
- Kane County: Wild Plants and Natural Areas, by Dick Young. 2nd ed. Kane County Forest Preserve District, 1994
- Peterson First Guides: Caterpillars. Trees. Shrubs.
- A Georgia Native Plant Guide, Tina M Samuels, Mercer University Press, 2005

Botanical Garden

Botanical gardens grow a wide variety of plants primarily categorized and documented for scientific purposes. Botanists and horticulturalists tend the flora and maintain the garden's library and herbarium of dried and documented plant material. Botanical gardens may also serve to entertain and educate the public, upon whom many depend for funding. However, not all botanical gardens are open to the public: for example the Chelsea Physic Garden. According to the International Agenda for Botanic Gardens in Conservation, "Botanic gardens are institutions holding documented collections of living plants for the purposes of scientific research, conservation, display and education."

Research: From the late 18th century onward, European botanical gardens began sending plant-collecting expeditions to va :ous parts of the world and publishing their findings. Voyages of exploration routinely included botanists for this purpose. Subsequent scientific work studied how these exotic plants might be adapted to grow in the garden's locale, how to classify them, and how to propagate rare or endangered species. The Royal Botanic Gardens in Kew, near London, has continuously published journals and more recently catalogues and databases since this time.

Educational Work: Educational projects at botanical gardens range from introductions to plants that thrive in different environments to practical advice for the home gardener. Many have plant shops, selling flower, herb, and vegetable seedlings suitable for transplantation. Some gardens such as the UBC Botanical Garden and Centre for Plant Research and the Chicago Botanic Garden have plant breeding programmes and introduce new plants to the horticultural trade.

Communal Garden: A communal garden (often used in the plural as communal gardens) is a shared garden for use by a number of residents. For example, the centre of many city squares and crescents (*e.g.*, in London) are maintained as communal gardens. Despite the name, such gardens are normally privately by jointly owned, with sharing of maintenance costs. Access may be restricted by locked gates, with keys available for residents, or only unlocked during daytime.

Community Garden: Community gardens are small plots of land allocated to groups of people by some organization that holds title or lease to the land, sometimes for rent, sometimes simply as a grant of land.

Community gardens run from 5'x5' plots to as much as 25'x25' plots. Usual sizes are in the 10' x 10' to 15' x 15' range. Community gardens are often run by a self-governing set of bylaws, some elect boards in a democratic fashion while others can be run by appointed officials. Most are run by a Non-profit organizations, such as a community gardening association, a church, or other land-owner. Others are run by a city's recreation or parks department, a school or University. There are many different organizational models in use for community gardens.

Community Gardens in the United Kingdom

The United Kingdom, community gardening is generally distinct from allotment gardening, though the distinction is sometimes blurred. Allotments are generally plots of land rented to individuals for their cultivation by local authorities or other public bodies-the upkeep of the land is usually the responsibility of the individual plot owners. Allotments tend (but not invariably) to be situated around the outskirts of built-up areas. Use of allotment areas as open space or play areas is generally discouraged.

The community garden movement is of more provenance than allotment gardening-many such gardens were built on bombed and derelict inner-city sites in the aftermath of The Blitz. A community garden in the UK tends to be situated in a built-up area and is typically run by an independent non-profit organisation (though this may be wholly or partly funded by public money).

It is also likely to perform a dual function as an open space or play area (in which role it may also be known as a 'city park') and-while it may offer plots to individual cultivators-the organisation that administers the garden will normally have a great deal of the responsibility for its planting, landscaping and upkeep. An example inner-city garden of this sort is Islington's Culpeper Community Garden, or Camden's Phoenix Garden.

Community Gardens in Australia

Australian City Farms & Community Gardens Network - national association of city farms, community gardens and sustainability education centres:

- The Rushall Garden in North Fitzroy, near Melbourne, Victoria.
- The Vera Street Community Garden in Toowong, Brisbane, Queensland.
- Veg Out Community Gardens in St Kilda, Victoria
- Ringwood Community Garden in Wantirna, part of Melbourne, Victoria

Community Gardens in the United States

In the United States, this term refers to gardens consisting of small plots (10ft by 10ft to 40ft by 40ft) leased to individuals by a community centre or municipality that retains ownership of the surrounding acreage. The British analogue is allotment gardening. In the last century, community gardens have appeared in inner-city neighbourhoods, small towns, and suburbs across the United States. In certain areas of the country, the legacy of progressive federal legislation funds small community gardening projects.

Community gardening in the U.S. contrasts to a related but distinct movement, Community-supported agriculture (CSA). Community gardeners are often prohibited by their leases from selling their produce commercially, although their gardens may provide fresh fruits and vegetables to local food pantries, cooperatives, and homeless members of their community.

In an interesting variant on the practice of reclaiming bombed-out areas for community gardens (also practiced during WWII in the ghettos of Eastern Europe), in American inner-cities, community groups have reclaimed abandoned or junked lots for garden plots. In these cases, groups have subsequently leased from a municipality that claims the property or claimed squatter's rights or a right to subsistence not currently recognized by the legal system. Two notable cases include the gardens of Manhattan's lower Eastside and the South Central Farm of Los Angeles, California.

Community Gardens in Boston, Massachusetts, USA

In the city of Boston, Massachusetts a series of 16 community gardens and pocket parks are owned and operated by the South End Lower Roxbury Open Space Land Trust.

Community Gardens in San Francisco, California, USA

In San Francisco, community gardens are created and governed in a variety of different ways. One group, a community-based and resident-led volunteer group in an underserved neighbourhood called Bayview Hunters Point, has created an open public garden on city-owned land, an enclosed food-producing garden, and many residential urban farms around privately owned homes. This group, the Quesada Gardens Initiative, is one of many organizations in the San Francisco Bay Area working at the nexus of environmental justice, health and wellness and food security, and community-building.

Community Gardens in Seattle, Washington, USA

The city of Seattle, Washington Department of Neighbourhoods supports a successful community gardening programme, called the P-Patch programme.

Cottage Garden

Originally, these gardens were created by the workers that lived in the cottages of the villages to provide them with food and herbs with flowers planted in for decoration. The more common flowers to the cottage garden were hollyhocks, delphinium and daisies in addition to the flowering herbs. The method of planting closely packed plants reduced the amount of weeding and watering required.

Today, a cottage garden is often primarily flowers and completely free-form in nature. Many gardeners attempt to use heirloom varieties of plants in their cottage gardens to preserve the antiquity of the method.

Cutting Garden

A functional garden used to grow flowers for indoor rather than outdoor display is known as a cutting garden. It is usually only a feature of large residences.

The cutting garden is typically placed in a fertile and sunlight position out of public view and is not artistically arranged, as it contains flowers for cutting. The cutting garden may comprise a herb garden and ornamental vegetables as well.

7

Care of Flowering Potted Plants

African violet: Outstanding among small, flowering potted plants, African violets in a good location should flower without interruption for years. In addition to their free-blooming character, they are well adapted to home conditions, easy to propagate and available in a wide range of flower colours. Miniatures as well as more compact standard varieties are usable in limited space.

Care of the New Plant:

- Maintain night temperatures between 65 and 70 degrees Fahrenheit; day temperatures 10 degrees warmer. Do not expose to temperatures below 60 or above 80 degrees Fahrenheit.
- Place in windows with bright light but no direct sunlight. Some midwinter sunlight is not harmful, but avoid it at other times of the year. If no suitable window space is available, plants grow well in fluorescent light. Place about 8 to 12 inches beneath cool white fluorescent tubes lit about 14 hours daily.
- Water frequently enough to keep soil moist, but allow slight drying between waterings. They are easily killed by excess soil moisture. Wick watering is adaptable.
- Humidity is important. In homes with low humidity, place on trays of gravel containing water. Home humidifiers can help.

Reblooming African Violets: African violets rebloom easily in the home. If located properly and watered regularly, little need be done besides occasional fertilization. Use either special African violet fertilizers or a houseplant fertilizer high in phosphorus. A very dilute fertilizer solution at each watering keeps growth constant and eliminates any chances of over fertilization.

Pale green leaf colour may indicate too much sunlight or low fertility. Do not use water softened by a system using salt in the process.

African violets seldom need pots larger than 4 inches in diameter. The danger of Over-watering and development of root and crown rots increases if pots are too big.

Old plants sometimes develop long woody stems. The tips of these plants may be cut off and rooted to form new, more compact plants. However, plants developed from leaf cuttings are generally more vigorous and bloom more abundantly.

Amaryllis: Amaryllis produce large, showy flowers before most of the foliage develops. Large bulbs produce two stems of flowers; smaller bulbs usually have only one. Bulbs ready for bloom or plants with bloom stalks started are available in early winter. With proper care these bulbs bloom yearly.

Care of the New Plant:

- When in bloom but before foliage develops, display in any desirable location indoors. However, as foliage develops, move plants to a bright window. Direct sunlight can shorten the life of the flowers, but after flowering is completed, give plants a bright, southern exposure with plenty of winter and spring sunlight.
- Keep bloom stalks from excessive stretching and to increase duration of bloom, keep plants in a cool room at night. When this is not possible, stakes may be necessary to keep the bloom stalk upright.
- Soil must be kept constantly moist while blooming. Drying will reduce flower life and cause the flower stalk to bend.

Reblooming Amaryllis: Despite its short flowering period, spectacular flowers rebloom easily in future years. Care is most important during spring and summer. After bloom is completed, remove blooms and stalks promptly so no seeds form. Do not damage or cut off any leaves. Place plants in direct sunlight. If the pot is small and the bulb large, transplanting may be necessary. The pot should be at least 2 inches larger in diameter than the bulb. Standard potting mixtures are quite satisfactory. When repotting, avoid as much root damage as possible. Repot before flowering but not after fully developed leaves are actively growing. The completion of flowering is the last time to replant.

Provide good soil drainage; never grow plants in pots without bottom drainage. Keep bulbs at temperatures between 60 and 70 degrees Fahrenheit. After danger of frost is past, potted bulbs may be moved outdoors. If garden space is available, sink pots to the rim. Position in sufficient sunlight where they can be easily watered.

Move gradually into a sunny location to reduce or avoid leaf burn. Keep plants actively growing during the summer. As danger of frost approaches, lift out pots containing the bulbs and store in a cool room (about 45 to 50 degrees Fahrenheit at night). Withhold water until the foliage dies. Light is not necessary. Bulbs require a two to three month rest period before growth and flowering begin again.

Flower buds should appear several weeks after moving plants to a warm temperature and watering is resumed. Bulbs that had four or more healthy leaves throughout the summer should be large enough to flower; those with less foliage may not flower. However, with proper care non-flowering bulbs can be redeveloped for blooming in future years.

Azalea: The popularity of potted azaleas has increased rapidly since they have become available throughout the year. Several types are used for forcing, but the care of all of them in the home is similar.

Care of the New Plant:

- Although azaleas can tolerate normal room temperatures, they will last much longer in cool conditions. Place the plant in the coolest part of the house where temperatures remain from 60 to 65 degrees Fahrenheit.
- Keep the plant in a well lighted location, but avoid full sun while in bloom. Sun directly on the flowers will fade

them and decrease their life. After flowering is finished, the plant should be moved to a sunny window.

- Never allow an azalea to dry out. Water the plant by submerging the pot in a bucket or pan of deep water. Remove it when bubbles stop rising. Allow excess water to drain out before returning the pot to the saucer. Never allow the plant to wilt.
- Use a complete, soluble liquid fertilizer about once every two weeks if you plan to keep the plant after flowering is finished.
- Keep faded flowers removed to prevent the development of disease problems and maintain a good appearance.

Reblooming Azaleas

Reblooming azaleas in the home is not a simple matter. After flowering has finished, remove withered flowers and keep the plant in a cool, sunny location. Shift the plant to a larger pot if it appears to be potbound. Azaleas need shifting only once every few years. It is unlikely that a newly purchased plant will need immediate repotting.

A good mixture for repotting can be made from three parts acid peat moss and one part soil. With careful fertilization and watering, azaleas can be grown in peat moss alone.

Submerge the potted plants outdoors in May in a semi-shaded, protected spot. Don't neglect it during the summer. It will need water and fertilization regularly. Watch for insect pests.

If foliage becomes yellow or chlorotic, iron uptake may be limited. This can mean root loss from too much or too little water. It can also result from a pH that is too high (soil too alkaline), or it may actually be a lack of iron in the soil. Apply a chelated iron or iron sulphate to correct the situation. Ammonium sulphate fertilizer also helps to lower the pH and maintain the acidity of the soil. Apply about one half teaspoon each of iron sulphate and ammonium sulphate to one quart of water. Apply this to the plants about every two to three weeks. About every third fertilization use a complete houseplant fertilizer instead of the above combination.

Before July 1, shape the plant by removing the tips of rapidly growing shoots. Keep the plant outdoors as long as possible in the fall. Protect it on the first cool or light frost nights, but bring it indoors when hard frost is likely. Place the plant in a sunny, cool room. It needs 40 to 50-degree temperatures from Nov. 1 to Jan. 1. Do not fertilize during this period and water only enough to keep the plant from wilting.

Buds should develop and swell. In January, move to a sunny window in a room where temperatures are near 60 degrees Fahrenheit at night. In a few weeks flowering should begin. If temperatures are excessively high during this period buds often develop poorly and new shoot growth begins. In warmer areas of the state, some of the Kurume varieties may be planted outdoors. None of the Indica, or Belgian, hybrids are hardy here.

Calceolaria

The oval, balloonlike flowers of calceolaria make its common name, "pocketbook plant," quite appropriate. It is usually available as a spring-flowering plant from about Valentine's Day through Easter.

Care of the new plant:

- These plants thrive in cool temperatures. For the longest display, locate plant in a cool room or window where temperatures are near 50 to 55 degrees Fahrenheit at night.
- Locate in bright light, but avoid direct sunlight, which may fade flowers and reduce their life span.
- Keep plants well watered at all times with slight drying between waterings. Overwatering easily damages their fine root system.

Care after Flowering: Calceolarias are annual plants grown from seeds. After flowering is completed, plants gradually decline and cannot be rebloomed successfully. Excessive heat or drying at any time makes them collapse quickly.

Chrysanthemums

Potted chrysanthemums are the most popular year-round flowering gift plant. They are available in a wide range of colours and forms. Although chrysanthemums normally flower in November, newly developed methods encourage flowering all seasons of the year. Potted mums last several weeks to a month when given proper care in the home.

Care of the New Plant:

- Place the plant in a window where it gets bright light but no direct sun.
- Check each day to make sure that the soil is moist. Do not allow the plant to wilt, but do not keep the pot sitting in water.
- Keep the plant at temperatures between 60 and 65 degrees Fahrenheit at night. Daytime temperatures may be 10 to 15 degrees higher.

Reblooming Chrysanthemums: Chrysanthemum varieties used for flowering potted plants are not usually suitable for Missouri gardens. However, if there is a desire to keep the plant, it may be rebloomed with proper attention.

Florists often pot several plants together. Separate these plants after flowering and repot them individually. Use a 5- to 6-inch pot. Keep these plants well watered and at temperatures close to 60 degrees Fahrenheit. When new shoots appear, move them to full sun and give plenty of water.

In summer, sink the pots outside in a sunny area where they can be given good care. Apply a complete liquid fertilizer about every two weeks. Additional plants can be started by rooting cuttings from the new growth.

To develop large, bushy plants, pinch out the tip of each new shoot when it becomes about 5 inches long. Remove only the soft shoot tips. Do not pinch the plants after Aug. 15.

Keep the plants outdoors as long as possible in fall. When light frost threatens, cover the plants or lift the pots and bring them indoors. They often can be returned outdoors during periods of mild fall weather. When cold, freezing temperatures persist, bring the plants indoors to stay. Place them in a sunny south window. Night temperatures around 60 degrees Fahrenheit are ideal. Day temperatures of 70 to 75 degrees Fahrenheit are best.

Chrysanthemums flower only in response to short days (long nights). Until the colour shows in the buds, plants must be kept where they will not be exposed to any light at night from a light fixture. Even dim light for short periods can disrupt flowering. When the buds begin to open, light at night will no longer effect them. Do not expect plants forced at home to be as perfect as those grown in a greenhouse. Greenhouse conditions are difficult to provide in the normal home.

Hardy garden mums also grow in pots. These are suitable for outdoor planting. Set them out in a sunny, well drained location in spring when the ground is warm. To keep these plants low and bushy, pinch out shoots when they are about 4 to 5 inches long. Discontinue pinching about mid-July.

Cineraria

Cineraria is a spectacular flowering plant for use on Easter and Mother's Day. Masses of daisylike flowers in either red, pink, blue, violet or white rise above a nosegay of large leaves. Cinerarias are seed propagated and grown at cool temperatures.

Care of the new plant:

- Plants last longest if kept at night temperatures between 45 and 50 degrees Fahrenheit with daytime temperatures between 55 and 65 degrees Fahrenheit.
- Constant soil moisture is critical for maintaining long life. They wilt quickly and can be seriously damaged if soil becomes dry. Water thoroughly to keep soil moisture constant but not too wet.
- Plants need good light but not direct sunlight, where higher temperatures may reduce flower durability. The heat, resulting in rapid soil drying, may also cause foliage wilting.

Care after flowering: With good care, cinerarias should be attractive for several weeks. These plants are grown from seed and are best handled as an annual. Once flowering is completed, they naturally decline and should be discarded.

Cyclamen

Cyclamen is available from Christmas through the spring holidays. Leaves are heart-shaped, often with attractive gray-green mottlings. Flower petals reflex back from the centers and are held on long stems above the foliage. Colours are pink, salmon, red and white.

Care of the new plant:

- Plants may flower for several months if kept at cool temperatures. Night temperatures close to 50 degrees Fahrenheit are best; however, a range from 50 to 60 degrees Fahrenheit is satisfactory. Day temperatures are less critical but should not run over 70 degrees Fahrenheit if possible. High temperatures may cause young flower buds to abort.
- During the winter, place plants where they get bright sunlight. At other times, place in bright light with little or no direct sunlight.
- Never allow plants to wilt. Water from the side of the pot, so water does not get into the centre of the "bulb." Keep soil moist with frequent watering, but do not allow water to stand in saucers beneath the pots.

Care after flowering: Cyclamen is a perennial plant and may rebloom a second season. After flowering, gradually withhold water until the foliage dies off. This begins the rest period. Do not water for six to eight weeks.

By midsummer, begin watering gradually. Repot if the bulblike corm is fairly large in the pot. When repotting, keep about half of the "bulb" above the soil line. At this time, a lightly shaded, cool spot outdoors, or a cool, shaded window indoors provides the best location.

As new leaves develop, resume normal watering and fertilization and move plants to a sunny location. Keep soil moist at all times, and feed with a houseplant fertilizer according to manufacturer's directions. Move plants indoors before danger of frost. Plants treated in this way should rebloom by midwinter.

If young leaves become stunted or curled, a pest known as cyclamen mite may be present. Although insecticides may combat this problem, infected plants are best discarded before the pest can spread to other plants.

Easter Lily

The Easter lily is a popular holiday plant that needs little special treatment. Well-tended plants should bloom successively for several weeks in the home.

Care of the new plant:

- Keep the plant moist, but be careful not to over-water. Check moisture daily and be sure the pot never stands in water. Root rots can be easily brought on by over-watering.
- Place the plant in a bright location, but avoid full sun.
- Keep the plant in a cool place and avoid drafts.
- When a new flower opens, carefully remove the yellow anthers. This will prevent pollen from smudging the petals.
- Cut off flowers as soon as they have collapsed.

Reblooming Easter Lilies

Outdoors: When all blossoms have faded, reduce watering so that the plant will gradually dry off. Cut off the stem a few inches above the soil after the top dries.

In May after danger of frost is past, plant the bulb in a protected place outdoors at a depth of 4 to 6 inches. Often the bulb will produce a few flowers again in late summer or early fall.

Easter lilies are not normally hardy in the Midwest but may persist in the garden for several years if placed in a protected spot and covered with a mulch.

Indoors: The forcing procedures seriously weaken the lily bulb. It is not practical to repot them and bring them indoors to try forcing them the following winter.

Gardenia

The glossy, deep green foliage and fragrant, waxy white flowers of gardenia make it a beautiful plant, but not an easy plant for home culture. Although available at other times, they are primarily potted plants used for the spring holidays.

Care of the new plant:

- Plants prefer temperatures close to 60 degrees Fahrenheit at night and 70 to 75 degrees Fahrenheit during the day. Maintain fairly uniform temperatures within these ranges.
- Maintain uniform soil moisture with good drainage. Fertilize with acid reaction fertilizers to keep soil slightly acid. If possible, use rain water, especially if tap water is "hard" or alkaline. Never use softened water because of its high sodium content.
- Place in sunny south or west windows.

Reblooming Gardenias: Unless located in a suitable environment, gardenias decline rapidly and should be discarded after flowering. Loss of flower buds is sometimes a problem with new plants. To help reduce bud drop, provide the best conditions possible. Misting over the plant while it becomes adjusted to a new location may help reduce bud drop.

When repotting is necessary, use a slightly acid soil mix high in peat moss. Repot in late spring after flowering is completed. If moved outside for the summer, they prefer a slightly shaded location. Most benefit from the addition of iron sulphate at three to four week intervals. Use 1/2 teaspoon to each quart of water. Soil acidity should be near pH 6.0 or slightly less. In early summer or at the time they are placed outdoors, prune back plants that have become too large.

Gloxinia

Care of the gloxinia is similar to that of the African violet, a close relative. Its large trumpet-shaped flowers rise above large, velvety leaves. Vivid colours include pink, red, white, blue and purple in solid colours as well as spotted or bordered with white. Gloxinias are most popular during the late winter and spring holidays.

Care of the new plant:

- Plants need warm nighttime temperatures (65 to 70 degrees Fahrenheit) along with fairly high humidity and good ventilation. Daytime temperatures should be about 10 degrees warmer. Placing plants on a tray of moist gravel or misting over new plants helps adjust them to home conditions.

- Gloxinias need more light than African violets, so place them in a bright window just beyond the reach of direct sunlight. In late spring, direct sunlight may burning the foliage and reduce the life of the flowers. If stems begin to stretch, light is inadequate.
- Keep soil uniformly moist. Plants can be damaged seriously if allowed to wilt. Small flower buds sometimes abort after plants are brought into the home. Either poor light, low humidity or excessive soil drying may lead to collapse of the young flower buds.

Care after flowering: Plants produce a fleshy tuber, which with proper care, may be rebloomed. After the plant has stopped flowering, gradually taper off watering until the leaves yellow and die. Place pot and tuber in a cool, dark location and stop watering. Give the tuber a rest period of eight to 10 weeks. As new growth appears, move the pot to a bright location and begin watering gradually. Tubers may be repotted at this time if necessary.

For spring bloom, do not water the tubers until after Christmas. Then remove the tuber from the old soil and plant shallowly into a new potting mixture. Place the top of the tuber above the soil line. Use only one tuber per pot. For starting plants, pots 4 or 5 inches in diameter are adequate. As shoots enlarge and leaves fully expand, begin normal watering and light fertilization. Plants should begin flowering in about four months.

Young plants may also be started from leaf or shoot cuttings. Techniques for starting plants from leaves are the same as for African violets.

Holiday Cacti

Thanksgiving cactus, Christmas cactus and Easter cactus are three tropical cacti that naturally flower near each of these holidays. Recent development of vigorous new hybrids, which flower between Thanksgiving and Christmas and in different colours and shades, has increased their recent popularity. In general, holiday cacti bear the stem characteristics of the Thanksgiving cactus. The Thanksgiving cactus has stem segments with saw-toothed margins; the Christmas cactus has rounded margins. Easter cactus is more intermediate in shape but develops more small spines at the junctions of the stem joints.

Care of the new plant:

- Cacti prefer a relatively cool night temperature between 55 and 60 degrees Fahrenheit. Night temperatures above 70 degrees Fahrenheit inhibit flowering and cause drop of existing buds.
- Place in some direct sunshine. Poor light promotes bud drop and poor growth after flowering. In summer, place plants in bright light but not direct sunlight.
- Water thoroughly, but allow to become moderately dry between waterings. Excess moisture results in root rot, especially during flower bud development in the fall during slow vegetative growth.

Reblooming Holiday Cacti

For flowering in future years, keep plants in a sunny window and fertilize regularly but lightly. Plants are epiphytic, that is, they grow in decaying organic matter rather than soil. When repotting, use sphagnum moss or a very loose mixture of organic materials. They do not need extremely large pots.

Keep in a bright window indoors during the summer, hang from branches of trees or set in other lightly shaded locations outdoors.

About mid-September, flower buds should begin to develop. Night temperatures of 55 degrees Fahrenheit result in the development of flower buds. When cool night temperatures are not possible, induce flowering by giving plants short, preferably nine-hour days, from about mid-September to mid-October. Cover with a light-tight box every evening for 15 hours of complete darkness. Flowering occurs about 21/2 months later.

Reduce watering and withhold fertilization during flower bud development and flowering stage.

Plants are relatively disease- and pest-free. However, overwatering or poor drainage of the potting medium may cause root rot and result in collapse of roots. Plants root easily from stem segments. To start new plants, break off the healthy older branches and root the tops in fresh, clean sphagnum moss. Water sparingly until rooting occurs and new growth begins. Wilted segments, if not too badly damaged, will recover and become firm as new roots develop.

Hydrangea

The potted hydrangea is popular for Easter and Mother's Day. With good care it should last several weeks in the home.

Care of the new plant:

- Water is the most important factor in caring for a potted hydrangea. The large foliage and flower clusters use large quantities of water. Never allow the plants to dry out. Check the soil several times a day. If you plan to be away for a day or two, cover the plant with a transparent plastic bag so that it will not lose water so fast. This should not be done frequently or it will weaken the plant and increase disease problems. Do not allow the pot to sit in a pan of water for long periods of time. If the plant wilts severely, submerge the soil and pot in a pan of deep water, and allow it to remain until the soil ball is thoroughly wetted. Drain and replace.
- Bright light will help to keep the plant in good condition. Place it near a sunny window but not where it gets direct sun which will fade or burn the flowers.
- Hydrangeas like cool temperatures. Move the plant to a cool room at night (50 to 60 degrees Fahrenheit).

Reblooming Hydrangeas

Outdoors: In warmer areas of Missouri, the potted hydrangea can be planted outdoors in a protected spot after flowering has finished and the weather has warmed up. After mild winters the plant may flower, but after severe winters it will often be killed back and will not bloom. Never prune hydrangea in late summer or fall. At that time flower buds are set, and pruning will remove buds that would produce flowers next year. Mulch the tops with straw or Styrofoam covers to protect from being killed during the winter.

Indoors: To rebloom the hydrangea indoors, cut the shoots back after the plant has finished flowering so that two nodes or pairs of leaves are left on each shoot. Repot in a mixture of equal parts of soil and peat moss. If soil is very heavy, add coarse sand or perlite. Grow the plant in full sun or a south window.

In May, move the plant outside and sink the pot where it gets full morning sun but light shade during the afternoon. Water regularly and fertilize with a complete liquid fertilizer about every two weeks. For extra large flower heads, allow only about three stems to develop. When removing extra shoots, take out those that grow toward the centre of the plant. Lift the pot occasionally to keep root growth from moving outside the pot.

To keep shoots from becoming too long, pinch back shoots during the summer. The last pinch should not be made any later than July.

Keep the plant outdoors as long as possible in the fall, but bring it indoors before a hard freeze. Keep the leaves on until November 1 if possible. Then pick off all the leaves by hand or put the plant in total darkness until all leaves drop naturally. The leafless plant must then be kept at temperatures from 35 to 40 degrees Fahrenheit for about six weeks. Keep the plant dry during this period.

After the cooling period, bring the plant to a sunny, cool room, with night temperatures of 55 to 60 degrees Fahrenheit. Water it well and fertilize about every two weeks. It should flower in about four months.

Colour: The colour of hydrangea can be controlled to some extent. Flowers are blue if grown in acid soil and pink if the soil is near neutral. The colour of white flowers cannot be changed by this method.

For blue flowers, use a soil mix with liberal quantities of acid peat added. Water the plant three times at 10 day intervals in late summer with aluminum sulphate (seven teaspoons per quart of water). After the cool dormant treatment, again add the aluminum sulphate solution about four more times at 10-day intervals. When fertilizing, do not use fertilizers that contain phosphorous.

To produce pink flowers, keep the soil more neutral (pH 6.5-7.0). If the soil is naturally acid, water with diluted lime water made with 2/3 teaspoon of hydrated lime in a quart of water. Use a complete fertilizer for regular fertilizations and add one teaspoon of superphosphate to each pot. If leaves turn yellow (chlorotic) due to this treatment, use chelated iron or iron sulphate to correct the condition. Use 3/4 teaspoon of iron sulphate per quart of water.

Kalanchoe

Compact varieties of kalanchoes in a greater range of flower colours have been developed recently. As a Christmas plant, the red-flowered types have been popular, but pink, orange and yellow varieties are available for other times of the year. The kalanchoe has thick, waxy leaves. Flowers are small, four-petaled, in abundant clusters held well above the leaves.

Care of the new plant:

- Plants prefer a night temperature from 60 to 65 degrees Fahrenheit. As with most houseplants, day temperature should be about 10 degrees higher.
- Locate the plant in a sunny south or west exposure during the winter. In the summer any bright location with a few hours of direct sunlight will be satisfactory.
- These plants can be damaged by over-watering. Allow soil to dry slightly between waterings. Fertilize with a common houseplant fertilizer at about monthly intervals.

Care after Flowering: With good care, kalanchoes may be rebloomed for another season. After flowering, shift the plant to a slightly larger pot. Use a well-drained, general-purpose potting soil. Cut back tall growth and old flower stems. Keep well watered in a sunny, warm window. Indoors, keep the plants in a sunny location, but after danger of frost move them outdoors for the summer. Gradually adjust them to outdoor conditions, so tender leaves will not burn. Keep in a bright, lightly shaded spot. New plants may also be started from cuttings taken in the spring or early summer.

Kalanchoes, like poinsettias, require short day lengths for flower bud development. Natural day lengths between October 1 and March 1 allow flower buds to form. During this time, keep the plant in a room where lights are not turned on during the natural dark hours. About six weeks of natural winter day lengths are required for flower buds to appear. After the flower buds are large enough to be seen above the foliage, day length is no longer critical. At this time, place plants in any location regardless of night lighting. Plants exposed to natural short day lengths in early October should begin flowering by January.

Poinsettia

The red flowering poinsettia is by far the most popular flowering potted plant for the Christmas season. White, pink and variegated white and pink are also available. Many new, long lasting varieties of poinsettias are now available. If properly cared for they may last a month or more after Christmas.

Care of the new plant:

- Poinsettias use a lot of water. Check daily. Make sure soil remains moist, but do not allow water to remain beneath the pot in the saucer or wrapping. Too much water will cause the roots to rot, and the plant will deteriorate.
- Keep the plant out of drafts. Excessively hot, dry air from heating ducts will reduce the life of the plant. Also avoid cold drafts. Poinsettias are semi-tropical and can not tolerate cold temperatures or rapid temperature changes.
- Keep the plant where temperatures remain above 60 degrees Fahrenheit. Temperatures between 65 and 70 degrees Fahrenheit are ideal. It is not necessary to move poinsettias into a cool room at night.
- Place the plant in good light but not direct sun.
- Do not allow the plant to wilt. Poinsettias are closely related to many desert plants. Their first response to dry conditions is to drop their leaves in order to cut down water loss.

Reblooming Poinsettias: **Poinsettias are perhaps the most difficult flowering potted plants to rebloom in the home. Unless there is a special interest to try the technique, discard the plant after flowering.**

After blooming, gradually withhold water. The leaves will then yellow and fall. Store the dried-off plant in a cool place with temperatures 50 to 60 degrees Fahrenheit until spring. Water only enough to keep the roots from drying out.

In spring, place the plant in a warm room and prune the stems back to about 6 inches. If there is more than one plant in the pot, divide and repot them at this time.

For repotting, use a well-drained soil. Houseplant potting soils available at garden shops are satisfactory. Or use one part garden soil, one part peat moss or leaf mold and one part sand or perlite.

After repotting, place the plants in a bright, sunny south window until frost danger is past. Sink the pot outdoors where it gets some wind protection but where it gets sun most of the day. Light shade in the hottest part of the summer afternoon is desirable. Lift the pot occasionally to keep roots from growing into the surrounding soil.

As new shoots develop, cut them back to allow two nodes or pairs of leaves to remain. Do not pinch back shoots after mid-August.

Plants may be started from cuttings, but rooting is fairly difficult under home conditions.

Keep the plant in good growing condition by watering and feeding regularly during the summer. Add a complete liquid fertilizer about once every two weeks.

Watch carefully for insect or disease problems and control immediately. Discard diseased plants.

Before the weather becomes cool in fall, bring the plant indoors and place it at a bright, sunny south window. Night temperatures for flower development should be between 60 and 62 degrees Fahrenheit. At higher temperatures flower development will be poor. Day temperatures may be 70 to 75 degrees Fahrenheit.

The poinsettia is a short day (long night) plant. Make sure that it receives no additional light at night while flowers are forming. This critical period begins about October 1 and continues until coloured bracts and flower buds are visible. Even short periods of dim light can prevent flowering. If the plant is kept in a lighted room, cover it every night at dusk with a light-tight bag or cover. Remove the cover at about 8 a.m. each morning.

If these procedures are followed carefully, the plants should flower by midwinter.

Potted Bulbs

Spring flowering bulbs often forced into bloom in pots are tulip, hyacinth, narcissus (daffodil), grape hyacinth and crocus. For maximum enjoyment, these plants are best purchased in the bud or early stages of bloom.

Care of the new plant:

- Place the plants in a cool spot where temperatures at night are about 60 degrees Fahrenheit.
- Do not place them in a sunny location. The warm temperatures in sunshine will speed flower development and shorten the life of the flowers.
- Keep the plants moist but not standing in water. Additional fertilization will not be necessary.

Reblooming Potted Bulbs: Do not try to rebloom indoors bulbs that have been forced. They can, however, be moved into the garden and will often flower the next spring. Discard any bulbs that have been forced in water such as hyacinths or paperwhite narcissus.

When the bulbs have finished flowering indoors, move the plant to a sunny location. Keep them wet enough so that leaves are able to mature and dry gradually. When the tops have died, remove the dry foliage and soil from around the bulbs. Store them in a dry place until fall, then plant them outdoors in the garden.

Spring Bulbs for Indoor Blooming: Bulbs made to flower at other than normal times are said to be "forced." The practice is commonly used to flower daffodils, hyacinths, tulips, crocus and other spring bulbs during the winter. With proper planning and care, bulbs can supply colour for the home from late November until early April when they begin to flower normally outdoors.

Forcing Bulbs in Soil: Steps for forcing include selecting most suitable varieties, potting bulbs properly in well-drained soil, providing proper cold temperature treatment, bringing bulbs into a cool room and moving the plants to a display location when well developed.

Any spring flowering bulb can be forced, but to be successful, select types and varieties known to be good forcers. Hyacinths and daffodils are generally the easiest. Tulips are slightly more difficult, but with proper treatments they can be forced successfully. Many small bulbs such as crocus, grape hyacinth and snowdrops also may be forced. The general procedure for forcing all these bulbs is similar. Purchase only top quality, flowering-size bulbs for indoor forcing. Good bulbs contain ingredients necessary for successfully producing roots, leaves and flowers.

Hyacinths: These are usually the most easily forced spring bulbs. They may be forced in water or potted. Potting should be done in late September or early October. Plants will flower about one month after being brought indoors from the chilling treatment. Earliest bloom from bulbs can be expected about mid-January. Specially prepared bulbs are available for forcing into bloom by Christmas. Specific directions are supplied with these bulbs if they are prepared for early forcing.

Lists hyacinth varieties most suitable for forcing along with the appropriate potting and forcing dates.

Hyacinth Varieties for Forcing

Variety		*Colour*	*Pot no later than*	*Earliest date to bring indoors*
Early forcing	Anne Marie	bright pink	Oct. 1	late December
	Delft Blue	porcelain blue	Oct. 1	late December
	Jan Bos	red	Oct. 1	late December
	L'Innocence	white	Oct. 1	late December

Contd...

	Variety	*Colour*	*Pot no later than*	*Earliest date to bring indoors*
Mid-season forcing	Carnegie	white	Oct. 1	early January
	Myosotis	pale blue	Oct. 1	early January
	Ostara	dark blue	Oct. 1	early January
	Pink Pearl	pink	Oct. 1	early January
Late forcing	City of Harlem	yellow	Oct. 15	mid-February
	King of the Blues	deep blue	Oct. 15	mid-February
	Lady Derby	rose pink	Oct. 15	late January
	Orange Boven	orange-salmon	Oct. 15	late January

Spring Flowering Bulbs—Daffodils.

The varieties most suitable for early forcing are Rembrandt, Cragford, Golden Harvest, King Alfred and Carlton.

Paperwhite narcissus are suitable for earliest forcing either in pots or in water.

Tulips Suggests tulip varieties suitable for forcing, but many others may be used.

Tulip Varieties for Forcing

Variety	*Colour*	*Earliest date to bring indoors*
Apeldoorn	orange-scarlet	early February
Apricot Beauty	salmon-rose	early January
Attilla	purple violet	mid-January
Bellona	golden yellow	early January
Blizzaard	creamy white	early February
Christmas Gold	deep yellow	early January
Christmas Marvel	cherry pink	early January
Couleur Cardinal	cardinal red	early February
DeWet (General)	orange	early January
First Lady	reddish-violet	early January
Garden Party	white-edged red	early February

Contd...

Variety	*Colour*	*Earliest date to bring indoors*
Gudoshnik	pale yellow streaked; rose pink	early February
Jewel of Spring	yellow-streaked red	early February
Kansas	white	early January
Make Up	white with red edge	early February
Merry Widow	red with white edge	mid-January
Olympic Flame	yellow-flamed red	early February
Orange Nassau	double-orange scarlet	mid-January
Orange Wonder	bronzy-orange	mid-January
Paul Richter	scarlet red	early January
Peach Blossom	double deep rose	early February
Preludlum	salmon with white base	early January
Queen of Sheba	mahogany-edged orange	early February
Westpoint	yellow	early February

All varieties should be potted before the end of October for adequate root growth and chilling. Dates listed in the table are the earliest that bulbs planted in October should be brought into forcing conditions. Most varieties flower about one month after being brought indoors, although there is variation due to varieties and individual forcing conditions.

Potting Stage

Materials: The following materials will be needed for potting bulbs:

- Pots 4 to 8 inches in diameter. Pots shorter than regular pots, known as azalea or bulb pots, should be used.
- Soil mix of one part garden soil, one part sand and one part peat. High fertility is not essential, but good drainage is important. No fertilizer is needed at potting time.
- Gravel or small stones for drainage.
- Labels of wood, plastic or metal.
- Thermometer for checking temperatures.

Procedure: All bulbs are normally potted in October.

- Place a 1-inch layer of gravel in the bottom of the pot to ensure good drainage.
- Add enough soil mixture to fill the pot so bulbs are placed as follows:

Daffodils: Plant so about one-half of the bulb will be above the soil line.

Hyacinths and Tulips: Allow only the tip of the bulb to show above the soil line.

Small bulbs (Crocus, Snowdrop, Grape hyacinth, etc.)

Plant so they will be about one inch below the soil line.

- Set the bulbs in the pot. One large bulb may be placed in each 4-inch pot. Use six tulips, three hyacinths, five daffodils or 15 crocus (or other small bulb) to each 6-inch pot.

Tulips should be placed with the flattened side of the bulb toward the outside of the pot. All bulbs in a pot should be of the same kind and variety to ensure uniform flowering.

- Fill around the bulbs with soil to the proper height and firm it with light pressure. Avoid tight packing of soil. After planting, the final soil line should be about 1/4 inch to 1/2 inch below the rim of the pot.
- Label each pot with variety of bulb, date of potting and expected date to begin forcing.
- Add water until it drips through the drainage hole in the bottom of the pot. After about one hour, dump out any remaining water in the saucer.

Rooting Stage

After potting, keep hardy bulbs where temperatures range between 35 and 48 degrees Fahrenheit — 40 degrees Fahrenheit is ideal. This may be a cool north room, basement, crawl space, bulb cellar, outdoor trench, refrigerator or cold frame.

An unheated basement or storage cellar is most convenient because temperatures don't fluctuate greatly and the cooling is quicker and more satisfactory. An old refrigerator may be used for a few bulbs.

A cold frame is easily built outdoors and may later be used for other gardening activities. Place the pots to be forced in the old frame and cover them with sawdust, straw, leaves, peat moss, shredded styrofoam or other material. The cold frame should be in a shady place or on the north side of a building so the soil is as uniformly cool as possible in October and November. No sash is needed on the cold frame.

When preparing only a few pots, place them on the surface of the ground close to a building. Cover them with peat moss, leaves, straw or similar material and invert a box or bushel basket over them for protection.

Bulbs stored outdoors will normally get sufficient moisture from the soil around them after initial watering. Indoors they should be kept moist at all times. Over-watering, however, may cause bulb rot.

Roots should develop soon after potting. Excellent root growth is essential to good growth and flower formation. Potted bulbs should be placed outdoors at least three weeks before the first hard freeze is expected. This is an important period for good root development.

Top-growth Stage

A few varieties may be brought indoors after about 12 weeks of cooling, but most will require 13 or 14 weeks to develop the necessary roots and top growth. Indoor forcing requires three to four weeks. To extend the bloom period, remove potted bulbs from storage at weekly intervals.

If soil and tops are frozen when they are brought indoors, place them in a cool room (about 40 degrees Fahrenheit) for two or three days to thaw out slowly. Don't touch the plant tops when they are frozen. If plants are not frozen, bring them directly to a cool, bright window where temperatures range close to 60 degrees Fahrenheit. Don't place them in direct sunlight. Keep bulbs watered, but fertilizer won't be needed.

Bloom Stage

When flower buds are almost fully developed, pots may be moved to the living room or other area in the house where they are to be displayed. Avoid placing them in full sunlight or close to heaters. The life of the flowers can be lengthened by moving the plants back to a cool room at night.

Bulbs that have been forced indoors are usually of little value for outdoor planting and should be discarded.

Forcing bulbs in water: Tender types of narcissus such as "Paperwhite" and "Soleil d'Or" don't require cold treatment before they may be forced into bloom. These are the most popular and dependable bulbs for forcing and may be grown in water with pebbles for support.

Hyacinths can be forced into bloom in containers that will support the bulb with only its base touching the water. Bulbs should be cleaned before placing them in glasses. The best time for starting the bulbs is in October.

The glasses containing the bulbs should be kept in a cool (45 to 50 degrees), dark location until tip growth is 3 to 4 inches long and the flower cluster emerges free from the bulb. This may take 8 to 12 weeks. When the top growth is well developed, move the glasses to a cool, bright window.

Starting Plants From Seeds: Plants of many annual flowers and vegetables may be started from seeds indoors. Vigorous plants started indoors flower sooner and produce an earlier harvest. Some, however, are best sown directly outdoors when weather conditions permit and are actually delayed by transplanting.

Growing plants from seeds can be a rewarding hobby and also allows home gardeners to grow varieties that may not be available from local plant growers.

Time for seeding indoors: The proper time for sowing seeds depends upon when plants may normally be moved outdoors. This period may range from four to 10 weeks, depending on the speed with which seedlings grow and the cultural conditions in the home.

Selecting Seeds for Planting:

- *Purity and Trueness to Type:* Good seeds should not contain seed of other crops or weeds and should be the correct variety. For best results, buy quality seed from a reliable dealer.

- *Packages and Storage:* Seeds sold in packages should show crop, variety, germination percentage and chemical seed treatments, if any. Seeds should be kept dry and cool to ensure good germination at planting. Laminated foil packets ensure dry storage. Paper packets are best kept in tightly closed cans or jars until seeds are planted.
- *Hybrid Seeds:* Many new vegetable and flower seeds are hybrids. Hybrid seed often costs more than seed of nonhybrid varieties. However, hybrids usually have increased vigour, better uniformity, better production and sometimes specific disease resistance or other unique cultural characteristics. Each gardener must decide whether the added benefit justifies the added cost. It often does.

Clean containers are extremely important for good germination and prevention of disease. These petunia seedlings are heavily seeded and will need prompt transplanting soon.

Containers for Sowing Seeds: Containers for starting seeds should be clean, sturdy and fit into the space available for growing plants in the home. The proper container helps get seedlings off to a good start and may save work in later stages of development.

- *Wood Flats, Fiber Trays, Plastic Trays:* Plants that are easy to transplant may be seeded directly in flats or trays for later transplanting into individual pots or wider spacing in flats. Starting seeds in such containers saves space as compared to seeding directly into individual pots. However, when time is more important than space, direct seeding in pots may be preferred.
- *Clay and Plastic Pots:* Both types can be cleaned and reused and provide excellent growth for transplants. They must be removed from the soil ball carefully at planting time. Seeds may also be planted directly into them.
- *Peat Pots:* These popular pots are made from peat or paper waste fibres and may be purchased individually or

in strips or blocks. They are porous and provide excellent drainage and air movement. The entire pot can be planted, so there is minimum root disturbance at planting time.

- *Compressed Peat Pellets:* Expandable peat pellets when dry are about the size of a silver dollar, but somewhat thicker. When placed in water they swell to form a cylindrical container filled with peat moss, ready for seeding or transplanting. They may be planted directly into the garden. Use the pellets in trays so that they are easily watered and held upright.

Soil Mixes and Other Growing Media: The medium used for starting seed should be loose, well drained and fine-textured. Prepared mixes are available commercially, or materials can be mixed at home.

- *Vermiculite:* This material when used alone provides good seed germination. It is clean, and if not contaminated during handling, will not need sterilization. If other seeding mixes are used, it is useful for covering seeds since it does not form a crust and provides easy emergence for the seedlings. Vermiculite is available in several size grades. For seeding, a fine grade is best.
- *Synthetic Mixtures:* Mixes that contain no soil are available for growing seeds. These contain either a combination of peat moss and vermiculite or peat moss and perlite. They may be purchased ready-made or can be mixed at home. To prepare such a mix, use 4 quarts of vermiculite, 4 quarts of peat moss, 1 tablespoon of superphosphate and 2 tablespoons of pulverized limestone. These mixes, as well as vermiculite used alone, have little fertility. Seedlings must be watered with a diluted fertilizer solution soon after they emerge.
- *Soil-vermiculite Mix:* Seeds may be started in a mixture of about one-third good garden soil and two-thirds vermiculite. Since good soil contains some fertility, prompt fertilization after germination is not so essential. When garden soil is used, the mix must be sterilized before seeds are planted in it.

- *Soil-peat-sand-mix:* Large seeds or vigorous-growing seeds may be planted in a mix of two parts good garden soil, one part shredded peat moss, and two parts either perlite, vermiculite or sharp sand. Soil sterilization is necessary.
- *Milled Sphagnum Moss:* This is a ground sphagnum moss sometimes used for starting seeds since it appears to have an ability to inhibit the seedling disease "damping off." It should be well moistened before use. Since it contains no fertility, prompt fertilization is essential after seeds have germinated.
- *Layered Mixes:* Another technique used for germinating seeds is to partially fill a flat or pot with a sterilized soil mix and then top it with a layer of vermiculite or milled sphagnum moss in which the seeds are planted. After germination, roots of seedlings move from this top area into the soil mix, which provides fertility, so adding liquid fertilizer is less critical.

Sterilizing Mixes and Containers: To guard against plant diseases carried in soil and on containers, cleanliness and sterilization of materials is important. Place the moist, but not wet, soil mix in a container that can be covered to keep the soil from drying rapidly.

If a cover is not available, cover with aluminum foil, and seal at the edges. The mixes may be sterilized directly in the pots or flats in which they are to be used. In addition to killing disease organisms, many weed seeds are also killed by sterilization.

Pinch a small hole in the centre of the foil and insert the bulb end of a meat or candy thermometer into the soil so that the bulb is about at the centre of the soil mass. Place the pan in an oven at 200 to 250 degrees Fahrenheit. Keep the soil in the oven until the thermometer shows a temperature of 160 to 180 degrees Fahrenheit.

Remove the pan and allow it to cool. Baking soil will give off a strong odour, so some ventilation may be desirable. The length of time necessary for sterilization depends on the volume of soil, as well as its moisture content. Dry soil cannot be sterilized well.

After the soil has been sterilized, make sure that containers, tools and working area are also clean and sterile. Clean soil can be easily reinfected by careless techniques.

In sterile soil, reintroduced disease may spread faster than it would in unsterilized soil

- *Sterilizing Containers:* Wash thoroughly in soapy water all used containers to remove all debris. Don't put wooden flats or plastic containers in the oven. Clean wood and plastic items by rinsing them in a solution of one part chlorine bleach and 10 parts water. Let them dry before filling with soil.

Steps to Seeding:

- Fill the container up to 3/4 inch from the top with the mixture to be used. It should be moistened before filling if the mixture is dry. Make sure the container has adequate drainage. Before filling, cover holes or cracks in the container with sphagnum moss or broken crockery.
- Level and gently firm the planting medium. A clean, small board may be used for leveling and firming.
- Using a narrow board or large wooden label, make shallow rows about 1 to 2 inches apart in the flat. When different seeds are used in the same container, they are easier to keep track of if planted in rows. If only one type of seed is used, it may be scattered or "broadcast" over the surface.
- Sow the seeds uniformly and thinly in the rows. Many small round seeds may be slowly dropped in the rows by tapping the package as it is held over the rows. Label each row promptly with plant type, variety and date of planting.

Large-seeded vegetables such as cucumber, cantaloupe and watermelon should be planted directly into peat pots. Other seeds may also be handled this way to save transplanting, but sowing is difficult with very small seeds. Some small seeds are now available in "pelletized form" to make handling easier. Plant two seeds per pot and later thin to one plant. This saves later transplanting and means less root damage at planting time.

- Cover the seeds with dry vermiculite or milled sphagnum moss. The depth of covering depends on the size of the seeds. Very fine seeds such as petunia or begonia need not be covered. Moisten the surface with a fine mist, or place the container in a pan of warm water to absorb it from the base.

  Don't place in water that is deep enough to run over the top of the pot or flat. It may mix seeds or cover them too deeply. As a general rule, seeds other than the very fine seeds should be covered about two times their diameter.
- Slip the flats or containers into clear plastic bags. Since they retain moisture, no additional watering should be necessary until after seeds have germinated.
- Place seeds in a warm location for germination. Generally, a range from 65 to 75 degrees Fahrenheit is best. A few plants such as larkspur, snapdragon, sweet pea, cabbage, broccoli and cauliflower are best started at about 55 degrees Fahrenheit. Don't place covered containers in direct sunlight.
- *Watch Daily for Germination:* Containers should be moved to bright light and the plastic bags removed as soon as germination is well under way. If not all seeds germinate at the same time, cut strips of plastic or cloth, and keep ungerminated rows covered until seedlings appear. Seeds are quickly killed if allowed to dry during germination. Watch closely for development of any disease and control promptly. After germination, those plants listed as preferring cool temperatures should be placed in a cool location.

Guide to Sowing Vegetable Seeds in the Home

Vegetables	*Time for seeding*[1]	*Comments*
Cool-season crops (broccoli, cabbage, cauliflower, head lettuce)	Late February	Grow cool, tolerate light frost outdoors after hardening
Warm-season crops (tomato, eggplant, pepper)	Late March	Keep warm. Do not subject to frost
Vine crops (cucumber, cantaloupe, squash, watermelon)	Late April	Sow directly in peat pots. Keep warm at all times.

Guide to Sowing Common Annual Flower Seeds in the Home

Time of seeding	Plant types	Germination time	Growth rate	Cold and frost tolerance after hardening
Early to mid-January	Begonia	10 to 12 days	Slow	None
Early to mid-January	Pansy	6 to 10 days	Medium	Good
Early to mid-January	Viola	6 to 10 days	Medium	Good
Early February	Lobelia	15 to 20 days	Slow	None
Early February	Stocks	10 to 14 days	Medium	Good
Mid-February	Black-eyed Susan vine	10 to 12 days	Slow - medium	None
Mid-February	Impatiens	15 to 18 days	Medium	None
Mid-February	Torenia	10 to 15 days	Medium	Medium
Late February	Petunia	6 to 12 days	Slow - medium	Slight
Early March	Ageratum	5 to 8 days	Medium	None
Early March	Scabiosa	8 to 12 days	Medium	Slight
Early March	Snapdragon	7 to 12 days	Medium	Medium
Early March	Verbena	12 to 20 days	Medium	Slight
Mid-March	Bells of Ireland	21 days	Medium	Medium
Mid-March	Dianthus	5 to 7 days	Medium	Medium
Mid-March	Salpiglossis (Painted tongue)	8 to 10 days	Medium	None
Mid-March	Vinca (Periwinkle)	10 to 15 days	Medium	None
Mid-March	Scarlet sage (Salvia)	12 to 15 days	Medium	None

When seedlings fall over at the ground line, they are being attacked by a fungus disease known as "damping off." If only a few seedlings are attacked, dig out and discard the infected plants and soil. Drench the entire soil mass with a fungicide if the disease is scattered throughout the flat or pot. This may not provide complete control. High temperature, poor light or excess moisture stimulate spread of the disease by weakening plants to make them more susceptible to it. Best control is cleanliness and prompt action when the disease appears.

Growing Seedlings

After seeds have germinated, they must be promptly given the best possible growing conditions to ensure stocky vigorous plants for outdoor planting. Cultural requirements must be considered carefully.

- *Light:* Seedlings must receive bright light promptly after germination. Place them in a bright south window if possible. If a large, bright window is not available, place the flats under fluorescent lights. A fixture containing two 40-watt fluorescent tubes is adequate. Place the seedlings about 6 inches from the tubes and keep lights on for 14 to 16 hours each day. As seedlings grow, the lights may need to be raised to prevent leaf burn as seedlings touch the tubes. Plants need some red and infrared radiation. Since this is not supplied by common fluorescent tubes, additional light from incandescent lamps or windows is necessary. If this cannot be given, use a fluorescent tube specially designed for plant growing. These are available under a variety of trade names.
- *Temperature:* Most annual plants and vegetables prefer night temperatures between 60 and 65 degrees Fahrenheit. Day temperatures may run about 10 degrees higher. If temperatures are warmer than this, leggy plants result. Cool-season vegetable crops and a few flowers already listed prefer night temperatures no higher than 55 degrees Fahrenheit and day temperatures near 65 degrees Fahrenheit. An unused bedroom, basement or sun porch is often a good location.

- *Moisture:* Good humidity is an asset for producing good plants. A humidifier may be used, or shallow pans of gravel filled with water may be placed as close to the growing area as possible. Flats should be kept moist at all times but never soggy. Allow drying between waterings, but don't allow seedlings to wilt at any time.
- *Fertilization:* Seedlings will need some fertilization for best development. Those in totally artificial mixes need prompt and regular fertilization. Use a soluble houseplant fertilizer as sold in garden centers, nurseries or on plant supply counters. Young, tender seedlings are easily damaged by too much fertilizer. Apply fertilizer at about half the recommended strength a few days after seedlings have germinated. After that, fertilize at two-week intervals with the dilution recommended by the manufacturer. Water and fertilize carefully.

Transplanting: As soon as seedlings have developed at least one set of true leaves and are large enough to handle, they should be transplanted to individual pots or spaced out in flats. Failure to transplant promptly results in crowded, spindly seedlings that may not be able to develop properly. If a hotbed is available, seedlings may be transferred directly to it. Artificial mixes may be used or a soil mix of one part soil, one part peat and one part sand works well for transplants.

To transplant, carefully dig up the small plants with a knife, spatula or wooden label. Let the group of seedlings fall apart and pick out individual plants. Occasionally if seedlings have been too close, they are difficult to separate. Gently ease them apart in small groups, which will make it easier to separate individual plants. Avoid tearing roots in the process. Handle small seedlings by their leaves; small thin stems break easily.

Poke a hole in the soil into which the seedling will be planted. Make it deep enough so the seedling can be put at the same depth it was growing in the seed flat. Small plants or slow growers may be placed 1 inch apart. Rapid-growing, large seedlings should be planted about 2 inches apart. After planting, firm the soil and water gently. If seeds were sown in individual peat pots or pellets, thin them to one seedling at this time.

Keep newly transplanted seedlings in the shade for a few days, or place them under fluorescent lights. Keep them away from heat sources. Continue watering and fertilizing as was done in the seed flats.

Vegetables easily transplanted include broccoli, cabbage, brussels sprouts, lettuce and tomato. Those with a little slower root development include cauliflower, celery, eggplant, onion and pepper. They may be successfully transplanted, however. Plants that do not transplant well and therefore are seeded in individual pots include cucumber, muskmelon, squash and watermelon. Most flowers normally grown indoors transplant well, but a few that are difficult to transplant include poppy, larkspur, lupine, sweet pea and cornflower. These are generally seeded outdoors, but to start them indoors, place them directly into individual pots.

Moving Plants Outdoors

- *Hardening:* Plants that have been growing indoors can't be planted abruptly into the garden without injury. To prevent any damage, they should be "hardened" before planting outdoors.

This process should be started at least two weeks before planting in the garden. If possible, plants should be moved to cooler temperatures outdoors in a shady location. A coldframe is excellent for this purpose. When first put outdoors, keep in the shade, but gradually move plants into sunlight for short periods each day. Gradually increase the length of exposure. Don't put tender seedlings outdoors on windy days or when temperatures are below 45 degrees Fahrenheit. Reduce the frequency of watering to slow growth but don't allow plants to wilt. Even cold-hardy plants such as cabbage and pansy will be hurt if exposed to freezing temperatures before they have been hardened. After proper hardening, however, they can be planted outdoors, and light frosts will not damage them.

- *Planting into the Garden:* When plants have grown large enough to handle easily and hardening is complete, they may be planted into the garden if weather conditions permit.

Carefully remove plants from the growing flats, leaving as much soil mix as possible around the roots. Dig the hole about twice as large as the soil mass around the roots. Set the plants at about the same level they have been growing in the pots. A few plants such as tomato and marigold are able to develop roots along the stem.

If they have become leggy, they may be planted deeper than they were previously growing. Place soil loosely around the roots and apply about 1 cup of a starter solution. This solution is made by dissolving 1 tablespoon of high-phosphorus fertilizer in 1 gallon of water. A 10-52-17 or similar analysis is satisfactory.

Plants grown in clay and plastic pots must be removed from them before planting. Those grown in peat pots or peat pellets can be planted intact. Breaking the base of the peat pot often helps improve root penetration and drainage. Make sure the top edges of the pot are thoroughly covered. If not covered, it may act as a wick and evaporate moisture around the root ball. This delays root penetration or even causes the plant to dry up on hot, sunny days.

Transplant on cloudy days if possible. In warm, sunny weather, cover the newly planted seedlings with newspaper tents or some other type of shading for two or three days until they are well established. Keep plants watered as necessary.

8

Lawn Establishment and Renovation

Establishment of turfgrass brings beauty to any landscape. While preparing an area for turf, do not hurry the process. Mistakes made at that time will be evident later and will cost extra in time and labour. The success of a beautiful turf is dependent on many factors, including initial soil preparation.

General Steps to Turfgrass Establishment:

- Obtain a soil fertility test and fertilizer recommendations
- Rough grade
- Apply lime if needed
- Apply fertilizer as recommended by soil test
- Apply organic amendments if needed
- Till materials listed above into top 4-6 inches of soil
- Perform finish grading
- Apply starter fertilizer and work into top inch of soil
- Apply seed
- Rake or drag to cover seed lightly
- Roll lightly
- Mulch
- Water
- Mow
- Control weeds

Preparation of the Site: You may not need to follow all of the above steps to establish your lawn. For example, the soil need not be completely tilled where grading and soil amendments are not required and when soils are not excessively compacted.

Once soil has been tilled, considerable time and effort are required to prepare a firm, granular seedbed. Where surface grade and soil physical conditions are acceptable, lawns can be re-established with minimal effort by killing the existing undesirable vegetation and incorporating seed into the surface. (See lawn renovation section.)

Regardless of how the site is prepared, it is important that seed be incorporated into the top 1/4 inch of soil. On loose, bare soils, this can easily be done by lightly raking the seed into the surface. On hard compact soils or soils with existing vegetation and thatch, the seed should be mechanically incorporated into the soil by verticutting, slit-seeding, or intense coring. Scattering seed on the surface without incorporation is a waste of time and money.

Soil Testing: A soil test is needed to determine the lime and fertilizer requirements to ensure good turfgrass establishment and future growth. The test results provide your soil pH value and if lime will be required. A soil pH between 6.0 and 6.8 is optimum for turfgrass growth. When pH is optimum, other nutrients such as nitrogen, phosphorus and potassium are readily available from the soil for plant uptake. Lime is used to raise soil pH, and sulphur is added to lower pH. Phosphorus and potassium needs, if any, will also be indicated on the soil test report. Phosphorus is especially important for root development and seedling establishment. Phosphorus is most effective when incorporated into the top 4 to 6 inches of the soil because it moves downward slowly.

Soil testing and sample bags may be obtained from your local MU Extension centre for a nominal fee. Send soil for analysis two months before your planting date to allow sufficient time for you to receive fertilizer recommendations and purchase the appropriate materials.

Soil Preparation: Proper attention to grading for surface drainage and conserving or developing topsoil will lead to easier care of your lawn in the years to come. Push aside existing topsoil when construction or excessive grade changes are required. Topsoil will be evenly spread over the site once the rough grading is completed.

When building a new home where a basement or foundation is dug, insist that the excavated subsoil is not combined with the existing topsoil. Trees, shrubs, groundcovers and lawns never develop to their full potential in poor quality subsoil. The initial investment in retaining on-site topsoil or amending poor quality soils will avoid the agonizing replacement of plants that could not tolerate the poor soil environment.

The area should be rough graded with gentle slopes to adequately drain or divert surface water without erosion. Slopes with a 1-foot drop in 50 feet should be adequate, but should not exceed a 1-foot drop in 15 feet. The finished grade should slope away from the foundation.

On sites where topsoil is added over subsoil, be sure that the finished grade is at least 3 inches below the top of the foundation of basement homes. Surface runoff can cause basement flooding if water runs over the basement foundation. Remove all debris such as large stones, tree roots and discarded construction materials that will interfere with turfgrass root growth and water movement through the soil.

Liming and Fertilizing: Apply the required amount of lime, phosphorus and potassium recommended on the soil test report for establishing a lawn. Where a soil test is not available, and if soils have required liming in the past but have not been limed for at least two years, apply 25 pounds of finely ground limestone per 1,000 square feet. Also apply 5 to 7 pounds of 0-46-0 phosphorus fertilizer per 1,000 square feet. Phosphorus is not very mobile in the soil and should be incorporated in the root zone (top 4 to 6 inches of soil) during this step. Nitrogen and potassium are highly mobile in the soil and easily can be distributed by surface application.

Thoroughly till the lime and fertilizer materials into the surface 4 to 6 inches of soil. Do not exceed 3 pounds of 0-46-0 fertilizer per 1,000 square feet if nutrients are not to be incorporated into the top 4 to 6 inches of soil. These practices are aimed at building the basic soil fertility to an acceptable level.

Broadcast and work into the top inch of soil 5 pounds of 20-27-5 or 19-25-5 fertilizer or 10 pounds of a 10-24-18 fertilizer, or the equivalent, per 1,000 square feet.

The fertilizer should be turf grade and contain 30 percent or more of the total nitrogen as water-insoluble or controlled-release nitrogen. Application at the recommended rate should provide adequate nutrition for the first three months of growth.

As an alternative starter fertilizer, apply 10 pounds of 10-10-10 or 12-12-12 farm-grade (soluble nitrogen) fertilizer or equivalent per 1,000 square feet (for example, 5 to 6 pounds of 20-20-20 fertilizer per 1,000 square feet). Using nitrogen fertilizers that contain only soluble nitrogen will necessitate additional nitrogen after six to eight weeks of growing weather.

Soil Amendments: A soil test will indicate if organic matter is required. Organic matter, such as peat moss or compost, can be added to clay soils to improve drainage and aerification. The same organic matter can be added to sandy soils to help hold water and nutrients in the soil. Fresh organic matter—manure, straw or fresh saw dust—is not usually recommended because it can cause a temporary nitrogen imbalance that will lead to slow growth and grass yellowing.

Reed-sedge peat and moss peat are also good sources of organic matter that tend to last longer in the soil than most other organic matter sources. Organic matter should be worked into the top 2 to 4 inches of soil before applying the starter fertilizer.

Final Grading: Rake the area to the finish grade just before seeding. Light rolling will indicate any low spots or other irregularities to the area. A proper final surface will be firm enough to prevent ruts made by seeding equipment but will be loose and crumbly so that seed easily can be raked into the top 1/4 inch of soil.

Once these steps have been properly followed, you are ready to install a turf by seeding or sodding.

Installation of Turf

When to Establish: Selecting the right time of the year to seed cool-season grasses is the most important factor in successful lawn establishment. The best time to seed cool-season grasses is between August 25 and October 1. Lawns seeded within a week of Labour Day are more likely to fill in completely by winter and produce a thicker appearance the following spring than lawns seeded in October.

Seeding in late summer is preferred because temperatures are still warm enough to promote rapid germination when provided with rain or irrigation. The cooler temperatures and shorter days of the approaching fall are ideal for further growth and development of young seedling grasses.

Establishing cool-season grasses from seed in the spring can be extremely difficult. In some cases, complete failure should be expected. Rainfall or watering required to germinate turfgrass seeds during spring establishment will also promote excessive weed growth, especially crabgrass. If you are lucky enough to beat the crabgrass, expect to continue the battle with frequent summer irrigations to prevent moisture stress. Even if moisture is adequate, summer temperatures may severely thin or completely kill seedling turf.

The constant summer watering required by seedling turf will also increase the chance of Pythium damping off and brown patch. Both of these fungal diseases are encouraged by the same conditions that promote seedling growth — wet soils, frequent watering and summer nitrogen.

Sod of Kentucky bluegrass or tall fescue can be installed during most of the year except in mid-winter when the ground is frozen. When extreme heat and drought conditions exist in summer, sod should be cut and laid during a 12- to 24-hour period. If done under drought conditions, the turf must be kept moist and cool. The soil should be watered enough to cool it before installation. It should be thoroughly watered again immediately after the sod is laid.

Seed vs. Sod: A quality lawn containing the recommended mixtures of specific grass varieties and species can be established with either seed or sod. When seeding, however, there are many more species and varieties from which to select. Most sod grown in Missouri is straight Kentucky bluegrass or a mixture of mostly tall fescue combined with a small amount of Kentucky bluegrass. Some tall fescue sod is also available on plastic netting.

Initially, seed is less expensive than sod. However, successful establishment is more risky with seed than with sod, and if re-seeding of certain areas or even an entire lawn is necessary, the overall expense may be less with sod. Also, the area is exposed to erosion because of the time required for seed to germinate and become well rooted in the soil. Sodding practically eliminates such problems, a consideration that may be especially important on steep hills or banks. Sod also reduces the chance of pesticide and nutrient contamination from surface runoff.

Sodding provides an immediately pleasing turf that is quickly functional, and it will compete with viable weed seed already present in the soil. Seeding usually requires weed control during the first year of establishment.

Seed should be used to establish lawns only in early fall or early spring, whereas sod offers less time limitation in that it may be established in nearly any season. Sodding of cool-season grasses in the spring is preferred to seeding.

Seeding: Before seeding, be sure that the soil has been prepared properly and is smooth and level.

Seed small areas by hand. A fertilizer spreader can be used in larger areas. When seeding by hand, try to apply five to 10 seeds per square inch. Heavier seedings will cause weak, spindly seedlings and spotty establishment. Calibrate spreaders to supply half the amount of seed in one pass over the area. Divide the amount of seed to be sown into two groups. Spread the first group of seed. Then spread the second group at a right angle to the first group.

Rake the seed to cover it with 1/8 to 1/4 inch of soil. Roll lightly to make good contact between seed and soil. Use a light layer of straw as mulch—one to one and one half bales per 1,000 square feet. This helps hasten germination, keeps soil moist and protects young seedlings. The soil should remain moist from the surface to just below the active root zone.

At first, this moist zone will be shallow and require light, frequent irrigation. Use a fine spray to sprinkle seeds one to four times per day until young seedlings are established. As the grass develops, irrigate deeper and less frequently. Always avoid puddles and runoff.

Sodding: Sodding is the installation of commercially grown turf. Sod has a carpetlike appearance consisting of green shoots attached to roots and soil. It usually comes in 3-foot sections, 18 inches wide, with less than 1/2 inch of soil attached.

Sodding has the advantage of almost immediate establishment, but its disadvantages are initial cost and the high amount of labour involved.

Choose high-quality sod that is actively growing. Sod is perishable and should not remain on the pallet or stack for more than a few hours. The presence of mildew and distinct yellowing of the leaves is usually good evidence of reduced turf vigour from being stacked too long.

To lay the sod, start with a straight edge such as a driveway or sidewalk. Unroll sod pieces tightly against each other, but don't overlap. Lay the pieces in a staggered, bricklike pattern. Using a sharp knife, cut pieces to fit curves or small areas. After the sod has been laid, roll it to ensure good contact with the soil. Be sure to water thoroughly, and water every day during dry, warm weather. Avoid soggy situations.

Watering: Newly seeded or sodded lawns require special irrigation. A newly seeded lawn requires daily watering and may need as many as four light waterings in a single day if conditions are dry and windy. Keep the seedbed moist, but not saturated, to a depth of 1 to 2 inches until germination occurs (green cast to lawn and seedlings 1/4 to 1/2 inch tall). At this stage it is crucial that seedlings not be stressed to the point of wilt.

Continue to water one to four times a day with light applications, approximately 1/8 to 1/4 inch of water per day. Straw applied at the time of seeding helps to shade the soil and to prevent rapid drying of the soil surface. It will also reduce soil erosion and seedling damage from the force of large sprinkler drops.

Watering with a light mist is best for establishing new lawns. As seedlings reach 2 inches, gradually reduce the frequency of watering and water more deeply. Begin mowing once a week when the seedlings reach a height of 3 inches. After the new lawn has been mowed two or three times, water deeply and infrequently. Newly sodded lawns may require watering one or two times a day. Begin irrigation immediately after laying sod. Plan your sodding operation so that a section of laid sod can immediately be watered while other areas are being sodded. Sod should be watered so that the sod strip is wet as well as the top 1 inch of soil below the sod.

The first irrigation requires about 1 inch of water to achieve complete wetting of the sod. After watering, lift up pieces of sod at several locations to determine if it has been adequately irrigated. Continue watering one to two times a day or as needed with light irrigations to prevent wilting and to ensure a moist soil just below the sod layer. As sod becomes established and roots penetrate and grow in the soil, gradually reduce the frequency of watering but wet the soil deeper. After sod has been mowed two or three times water deeply and infrequently.

During hot, windy conditions, establishing sod may require several light mistings per day to prevent wilt and potentially high lethal temperatures. In this case, mist the sod lightly just to wet the leaf surface and not to supply water to the soil. Misting cools the grass plant as water is evaporated from the leaves.

Do not over-irrigate the soil because that will inhibit sod roots from growing into the soil. In situations where daily watering is not possible, thoroughly water the sod and soil to a depth of 6 inches. This will delay the rooting time of sod but will reduce the chance of rapid drying and severe loss of grass.

Control Weeds

Spring Seeded Cool-season Grasses: Even though spring seeding of cool-season grasses is not as successful as fall seeding, spring and early summer seeding may be unavoidable. Some tips may help your attempt to establish cool-season grasses.

- *Seed as Early as Possible.* In Missouri, wet spring conditions may prevent using equipment, especially on construction sites. Prepare soil and spread seed in February or March if conditions are dry. Seeds will lay dormant until warm April temperatures cause germination.
- Consider using a slit seeder or drill that causes minimal surface disturbance. This reduces competition from weed seeds that may be brought to the surface with complete tillage. These types of seeders are available from rental services or lawn care professionals.
- Competition from crabgrass is a major problem with spring seeding, especially when the lawn is seeded with mostly Kentucky bluegrass. Siduron (Tupersan) is the only preemergent crabgrass herbicide that can be used with new spring seedings, and it will not harm the planted turfgrasses as they emerge. It should be applied as the final operation after seeding and before crabgrass germination. If lawn grasses are seeded in February or March, wait until early April to apply siduron so that crabgrass control will last longer into the summer. Shaded areas of the lawn do not need crabgrass control.
- After three or four mowings, usually at least a month after grass emerges, a label-recommended rate of a broad leaf herbicide on actively growing grass if broad-leaved weeds appear to be overpowering the grass. Do not use herbicides if you do not have a serious weed problem — frequent mowing and proper starter fertilizer may be the only management necessary to establish grass. Broad-leaved weeds can be easily controlled during the following year, provided the turf is able to initially establish without being crowded out by weeds.
- Use slightly higher seeding rates to ensure rapid cover and improve competition with annual weedy grasses. Perennial ryegrass, because of its rapid germination, stands the greatest chance of filling in during the first

year. Tall fescue is the next best choice for showing at least a modest stand of grass by the fall. Kentucky bluegrass requires two to four weeks to germinate in the spring. It usually develops a modest stand of grass in the spring. Never use more than 20 percent perennial ryegrass in a mixture.

Fall Seeded Cool-season Grasses: Fall seedings are much more successful than spring seedings because crabgrass and summer stress are not a problem. However, winter annual Broad-leaved weeds such as henbit, chickweed and speedwell may require control. These weeds germinate September through November, grow during the winter and result in a dense mat of weeds by early spring.

If weeds dominate the new lawn in the fall, use a postemergence herbicide after the lawn has been mowed at least three times, usually about 45 days after grass germination. Apply broad leaf herbicides on a warm sunny day in the fall when no frost is expected. If needed, winter annual Broad-leaved weeds can also be killed in late March and April. Even though these weeds naturally die in May, without herbicide application, they should be controlled in early spring if they are overly competitive. If annual Broad-leaved weeds are left uncontrolled, turf will appear thin and weakened through the summer in areas where weeds have naturally died.

Lawn Renovation

Lawns may occasionally thin out and become weedy as the result of poor management, pests or severe summer stress. Lawns that do not need regrading or soil tilling may be renovated by simply controlling weeds, planting seed and fertilizing. Also, determine the cause of lawn decline so that you can adjust your lawn care programme. Establishment practices mentioned above for time of seeding, weed control, fertility and watering also apply to renovation.

General Steps to Renovate Turfgrass:

- Control existing vegetation that you don't want. If you use a selective broad leaf herbicide (2,4-D, dicamba, or mecoprop), wait one month before the next step. Proceed after seven days following total vegetation control with Roundup herbicide and five days with Finale

- Set your mower as low as possible, scalp off all existing vegetation, and rake clean
- Prepare surface with power rake, verticutter or core aerifier. Skip steps 4, 6 and 7 if you use slit seeding equipment, which places seed directly into the soil
- Rake loosened thatch and existing debris
- Add fertilizer
- Make a final pass to create open channels for seed collection
- Apply seed in two directions
- Lightly rake seed into soil surface
- Water frequently until established

Renovating Tips

Assess Situation: First, assess your lawn situation for desirability and appropriateness of the grass, presence of weeds and bare spots. Kill existing vegetation (grasses and weeds) if any one of the following apply:

- Less than 50 percent of the area is in the desired turfgrass
- Only a cool season lawn grass is desired but bermudagrass has invaded. If bermudagrass is in localized patches or is encroaching from a neighbour along your property line, kill all vegetation only in those areas
- You want to use a turf-type tall fescue but currently have Kentucky bluegrass, or K-31 tall fescue
- You have several weeds that require non-selective control such as quackgrass, tall fescue, bermudagrass and nimblewill.

Lawns that have more than 50 percent desirable grass, but still appear thin and weak with several broad-leaved weeds, may need only control of broad-leaved weeds with broad leaf herbicides. After weed competition is reduced, lawns may regain their original density.

Bare patches less than 4 inches in diameter likely will fill in on their own with proper fertility and no additional seeding.

Larger areas that remain too thin a month after broad leaf herbicide treatment may be overseeded. Shaded areas of lawns with thin turf and exposed soil often appear to need additional seeding. Renovate these areas but do not kill existing turf that has thus far survived the difficult growing conditions of shade.

Prepare Surface for Seeding: One of the most important steps in renovation is placing the seed in contact with soil. This sounds simple, but most lawns have thatch - an intermingled layer of both dead and living plants - over the soil surface. This brown, decomposing layer may be up to 1-1/2 inches thick. Lawns with more than 1/2 inch of thatch should be dethatched.

Seed placed on or in the thatch layer may germinate, but the eventual stand of grass that develops will be poor. To ensure good seed-to-soil contact, use power equipment to prepare the surface for seeding. Power rakes, verticutters, slit seeders and core aerators are effective machines for properly preparing the surface for seeding.

Dethatching equipment is sometimes available from local rental stores. Local lawn care and landscape companies also provide dethatching and renovation service. Two to four passes over the lawn may be necessary depending on the soil hardness and thatch thickness. Raking of excess debris will be necessary as well.

Power Rake: Power rakes are lawn mower-like machines that have spring tines instead of blades. The power-driven spring tines rotate in a vertical position and pull the thatch to the surface.

Spring tines that attach to the lawn mower blade do not adequately prepare the surface for seeding. A half inch of thatch can be removed with a severe power raking, but be prepared to handle several bags of loosened thatch. Power rakes that can be adjusted so that the soil surface is slightly scratched will ensure good seed-to-soil contact but may not loosen hard soils.

Vertical Mower: Vertical mowers are similar to power rakes except that they have fixed or flail-type blades that cut groves through the thatch and into the soil. Fixed-blade vertical mowers, sometimes called slicers, are better at cutting into the soil. Heavy-duty, walk-behind vertical mowers are superior to power rakes for preparing a seedbed. They cut deeper into the soil, 1/2 to 3/4 inch, redistribute some soil onto the surface, and provide some soil loosening that helps seedling establishment.

Power rakes and vertical mowers can accumulate a large amount of grass debris on the surface that should be removed so that the seed can contact the soil. After sufficient disruption of the surface, lightly rake the dead debris off the surface, leaving the loosened soil behind. After raking, a final pass with the equipment will produce closely spaced groves that are excellent channels for catching broadcast seed.

Slit Seeder: Slit seeders are smaller versions of agricultural grain drills, but with closer spacings between discs. The discs slice open the ground just ahead of a small tube that drops the seed into the open furrow. A packing wheel follows the seed tube and closes the sliced ground.

Some slit seeders also have a vertical mowing attachment in front of the discs that loosens and redistributes soil on the surface. A slit seeder sows grasses in distinct rows 2 to 6 inches apart depending on the space between the drill discs. It is important to repeat the seeding process at different angles. This usually can be accomplished with two to four passes across the lawn making a checkerboard pattern.

Core Aerifier: Coring machines have a series of hollow tines that cut several cylindrical holes in the ground 2 to 3 inches deep and 1/2 inch in diameter. Some machines will not penetrate hard and dry compacted soils.

Coring machines are primarily used to provide a channel through which air, fertilizer and water can enter the soil. This channel cut through the thatch and into the soil helps to reduce soil compaction, increase root growth and redistribute soil on the upper surface of the thatch. The soil should be sufficiently moist so that the coring machine can penetrate. If a screwdriver can be pushed by hand 3 inches into the ground, the soil is about right. Soils that are too wet will clog the hollow tines and prevent core removal.

The soil cores should be left on the surface to create a more favourable environment for breakdown of the thatch layer by naturally occurring soil microbes. When the soil cores have dried, they can be chopped and dispersed over the surface with a mower, verticutter or power rake. If coring is to be the primary means of preparing a seedbed, the area should be cored to produce at least one hole every 2 inches. Seed and fertilize while the core holes are open.

Core aerifiers are of many different sizes and shapes. As with all equipment, check to see that it will fit through the gates on property gates before renting or contracting a service.

Home Lawn Weed Control

Bindweed: Creeping vine, leaf like arrowhead, flower white or pink; perennial.

Chickweeds:

Common: Light green delicate vine; winter annual.

Mouseear: dark green hairy leaves; perennial.

Goosegrass : Coarser than crabgrass, leaf base flat whitish; summer annual.

Ground ivy: Creeping stems mostly square, leaves mostly circular, perennial.

Henbit: Upright squarish stems, blue and lavender flowers; winter annual.

Knotweed : Tough viny stems, leaves small and single; spring germinating annual.

Mallow : Upright stem, five-point leaf; summer annual or biennial.

Nimblewill: Fine wiry stems, frail inconspicuous seeds, winter dormant; perennial.

Plantain, (broad leaf or Rugels): Thick oval leaves, long seedhead; perennial.

Plantain, (narrow or buckhorn): Long narrow leaf, seedhead atop wiry stem; perennial.

Shepherd's-purse : Young leaf like dandelion, triangular seed pod; winter or summer annual.

Spurge (prostrate or mat): Plant sap milky, stems often matted, summer annual.

Yellow Nutgrass (nutsedge) : Triangular stems, yellowish green; treat as perennial.

Note: The University of Missouri intends no endorsement of commercial products or trade names.

Weed-free turf beautifies and adds greater usefulness to home landscapes, recreational areas, and public and industrial grounds. Proper management is the most effective factor in weed control, but unfavourable climate, insects, diseases or abuse permits weeds to invade your lawn. Under such conditions, careful use of appropriate herbicides permits the turf to recover its original value.

Weed Control Principles

A dense stand of healthy grass provides the best weed control. Because most weeds are "opportunists" that invade weakened lawns, the fight against weeds starts with good turf management. Cultural practices such as mowing, fertilizing and watering should be done in a manner and time that will favour the grass rather than the weeds. Height of mowing influences competition against weeds such as crabgrass—the higher the cut, the lower the infestation. Frequent light sprinkling encourages shallow-rooted weeds and seed germination. Less frequent "deep-soak" watering that maintains a dry surface layer provides the grass with a competitive advantage.

Temperature, light, soil moisture and other factors determine the time and extent of weed germination and development. Some weeds germinate in early spring while others sprout in summer or fall. If conditions are favourable, a weed may be particularly abundant in a given year, but under different conditions the next year, it may be little in evidence.

Weeds that germinate in the spring from seed and mature in summer or fall each year (crabgrass and knotweed, for example), are designated as summer annuals. Chickweed, henbit and others germinate in fall or late winter and mature in late spring. These are called winter annuals.

Preemergence herbicides (chemicals applied before germination of weed seeds) must be applied in spring to control summer annuals, but in fall to control winter annuals. After weeds appear, postemergence herbicides must be used.

Perennial broad leaf weeds live more than two years but may produce seeds for new plants each year. Like biennials, which require two years to complete a life cycle, they store food in fleshy roots for next year's growth. Generally, postemergence herbicides are required for control of biennials and are more effective during the first year's growth. Fall often may be a good time to apply such herbicides.

Herbicide Application: Although most herbicides are formulated with reliable safety factors, application rates higher than those recommended may cause injury to turf and ornamental plants. Many people overapply herbicides, especially when using fertilizer-herbicide combinations. The user needs to follow instructions on containers carefully to avoid overdoses.

Preemergence treatments are applied before weeds sprout from seeds. Apply two weeks ahead of germination. Less effective control may be expected if applied more than a month before germination. Applications should not be made until excess lawn clippings and leaf litter are removed. Homeowners will find many over-the-counter preemergence products at local garden centers for control of summer annual weeds. Many are often referred to as "crabgrass preventor." These products should be applied by April 15 in the northern half of Missouri; by April 1 in the southern half. Be sure to follow label recommendations for application rates and irrigation requirements. Crabgrass preventors, for control of summer annual grasses, require irrigation or rainfall to activate.

Postemergence herbicides containing such materials as 2,4-D, MCPP, dicamba, triclopyr, MCPA and dithiopyr are applied after weeds appear. Liquid sprays can be more effective than dry materials, especially on hard-to-kill weeds. Apply postemergence materials when weeds are growing vigorously. Tough, old weeds are hard to kill, and if mature seeds are already formed, the lawn is likely to be infested again next year. Amine forms are safest because they give off fewer vapours that might damage other plants. Volatile ester formulations should not be used around ornamental plants. However, many granular products now have formulations that offer excellent control of target weeds. Select a time when winds are calm to prevent spray drift. Using granules impregnated with herbicides near ornamentals will minimize such hazards.

Fertilizer-herbicide combinations are extremely popular because they combine two operations. Combinations with preemergence chemicals are generally effective since both fertilizer and herbicide action are dependent on contact with the soil (requiring rainfall or irrigation). Postemergence herbicide action depends more on absorption by leaves, and granules in such combinations do not adhere well to smooth-surfaced leaves.

They will stick better if applied when weed leaves are damp, perhaps with morning dew. "Weed and feed" materials present a conflict in desirable actions. Proper time for weed control often does not coincide with the most desirable time and rates for fertilizing. If used for follow-up fertilizations, there is danger of herbicide overdose.

Equipment : Fertilizer spreaders can be used for applying granular herbicides. Be sure to adjust the spreader to apply recommended rates. If possible, apply half the desired rate in one direction and the remaining half at right angles to the first application.

Gravity flow applicators, compressed-air sprayers or types attached to a garden hose are effective for liquid applications. High pressures cause mists subject to drift and should be avoided. Sprinkler cans or sprinkler nozzles attached to a gallon container can be used on small areas.

A sprayer used for application of 2,4-D and related chemicals should not be used to spray garden or flower plants. Cleaning procedures are not always reliable. To be safe, have a separate sprayer for weed-killing purposes.

Eliminating weeds is of little value unless enough desirable grass is present to fill in bare spots. A re-seeding programme deserves first consideration if the turf is so weak that it will not recover once weeds are eliminated. Study soil and other conditions to determine reasons for low vigour of the original turf.

Note: The University of Missouri intends no endorsement.

Annual Broadleaf Weeds

Winter Annuals: Can be controlled with Gallery preemergence herbicide applied in late September or with a three-way premix postemergence herbicide applied in late fall or early spring on a warm day.

- Common chickweed
- Deadnettle
- Henbit
- Peppergrass
- Shepherd's-purse

Summer Annuals: Knotweed and speedwell germinate very early in the spring. Mallow is a biennial weed. Black medic and yellow wood sorrel can be perennials. Applications should be made early in the weeds' development for best control.

- Black medic
- Carpetweed
- Knotweed
- Kochia
- Lambsquarters
- Mallow
- Pigweed
- Puncturevine
- Purslane
- Ragweed
- Speedwell (Veronica)
- Spurge, spotted
- Vervain
- Yellow wood sorrel

Annual Broadleaf Herbicides

Trade Name/common Name

Preemergence

- Gallery/isoxaben

Postemergence

- Cool Power/MCPA + triclopyr + dicamba
- Dissolve/2,4-D + MCPP + dichlorprop
- Horsepower/MCPA + triclopyr + dicamba
- Power Zone/carfentrazone + MCPA + MCPP + dicamba
- Speed Zone/carfentrazone + 2,4-D ester + MCPP + dicamba
- Super Trimec/2,4-D + 2,4-DP + dicamba
- Three-Way/2,4-D + MCPP + dicamba
- Triamine/2,4-D + MCPP + dichlorprop
- Trimec Classic/2,4-D + MCPP + dicamba
- Triplet/2,4-D + MCPP + dicamba
- Tri-Power/MCPA + MCPP + dicamba
- Turflon II Amine/2,4-D + triclopyr (amine formulation)
- Turflon Ester/triclopyr ester
- Weed-B-Gon/2,4-D + MCPP + dicamba

Preemergence herbicides for annual grass control will provide preemergence control of many winter and summer annual broad leaf weeds. Apply in the spring for summer annuals; early fall for winter annuals. Please refer to the labels for information.

Perennial Broadleaf Weeds: Applications in the spring or fall while weeds are active will give control. Best control is achieved in the fall when weeds are translocating food reserves from tops to roots in preparation for winter.

- Bindweed
- Dandelion
- Docks
- Ground ivy
- Mouseear chickweed
- Plantains
- Poison ivy
- Red sorrel

- Thistles
- Violets
- White clover
- Yarrow

Perennial Broadleaf Herbicides

Trade name/common name

Postemergence

- Cool Power/MCPA + triclopyr + dicamba
- Dissolve/2,4-D + MCPP + dichlorprop
- Horsepower/MCPA + triclopyr + dicamba
- Power Zone/carfentrazone + MCPA + MCPP + dicamba
- Speed Zone/carfentrazone + 2,4-D ester + MCPP + dicamba
- Super Trimec/2,4-D + 2,4-DP + dicamba
- Three-Way/2,4-D + MCPP + dicamba
- Triamine/2,4-D + MCPP + dichlorprop
- Trimec Classic/2,4-D + MCPP + dicamba
- Triplet/2,4-D + MCPP + dicamba
- Tri-Power/MCPA + MCPP + dicamba
- Turflon Ester/triclopyr ester
- Weed-B-Gon MAX/triclopyr+ MCPP + dicamba

Advantages of fall applications of postemergence broad leaf herbicides :

- Less risk of injury to desired plants.
- Both perennial and winter annual seedlings can be controlled.
- Perennials weakened by herbicide may be killed by winter weather.
- Bare spots left by dead weeds will be filled sooner by bluegrass and other cool-season grasses.

The speed of an applicator pushing a spreader is not as critical to determine because the speed is automatically taken into account when setting the flow rate. However, because different applicators walk at different speeds, applicators should calibrate their own spreaders. Some products list spreader settings on the label. Many of these products require that the applicator walk at a specified speed (*e.g.*, Product M in Brand X spreader at Setting H applied at 3 mph). In that case, you will need to get some idea of your normal walking pace in miles per hour. With a spreader (in the closed position) half filled with granular material, measure the time required to walk a measured distance as described above.

On self-propelled or pull-behind sprayers, speed is critical for accurate spraying. Speedometer gauges or electronic LED devices on sprayers and tractors can give inaccurate readings because of wheel slippage, worn tires or improper tire pressures. Measure and mark a distance of 100 to 200 feet in level turf (avoid hills) similar to that being sprayed, and check the time required to travel the distance. Most accurate measurements will be made with the sprayer half full of water. Be sure the sprayer is moving at the desired speed when measuring the time it takes to travel the distance between the starting and ending markers.

The operator can select a gear and throttle setting that are safe and proper (maintaining minimum boom bounce) and determine what that speed is. The other option is to select a speed and through trial and error determine the gear and throttle setting required to achieve that specific speed. Keep in mind the throttle should be set to full on some equipment for proper operation of power takeoff equipment. Make several runs and measurements to determine your speed as accurately as possible.

Selecting Spray Tips: Hand-held and knapsack sprayers most often have a factory-installed nozzle. The applicator is then required to calibrate the sprayer with the factory tip. Some knapsack sprayers have interchangeable standard tips.

Spray tips come in numerous types, angles and materials. Flat fan spray, full or hollow cone and flood tips are the more common spray tips used in turfgrass spraying. A flat fan is the most widely used tip and under varying ranges of pressure is a good selection for many contact and systemic pesticides. Flat fans (specifically designed for low pressures) offer good spray drift control when operated at low pressures.

Spray tip material determines the life of the tip, depending on the solutions sprayed. Brass, nylon, stainless steel, hardened stainless steel and ceramic offer a selection of spray tip materials with performance ratings that range from poor to superior, respectively. Keep in mind that superior wear tolerance usually implies a higher initial price.

Spray angles of 65, 80 and 110 degrees are commonly used in turfgrasses. Depending on the nozzle type and size, the operating pressure can have a significant effect on spray angle and quality of spray distribution. Proper nozzle spacing and height of the boom help achieve optimum spray angles. See your manufacturer's recommendations for nozzle spacing and boom height.

When selecting a spray tip, find one that will deliver the desired spray volume and will be the most effective for the type of pesticide application you are making. Check pesticide labels for recommended spray volumes. Make sure that all spray tips in the spray boom are of the same pattern, angle and size. A damaged tip can cause irregular patterns in the spray and should be replaced. Damaged or worn tips will usually be located when checking the flow rate. If several tips are damaged or worn, replace all tips with new ones.

Pressure: The applicator is in control of pressure when using hand-held or knapsack sprayers. The applicator is required to pump the sprayer with air to maintain adequate spray pressures. The spray pressure on larger spray units is controlled by pump speed, throttling valve or pressure regulator. Pressures can be regulated up or down to get the desired flow rate. Higher pressure increases the flow rate through the nozzle, decreases spray particle size (higher potential for drift), and increases the rate of tip wear. Therefore, follow the manufacturer's pressure recommendations for tips selected.

When spray volume and tip selection are determined, the applicator is ready to start the sprayer and adjust the flow rate.

Calibrating Sprayers

Hand-held and knapsack sprayers: It is important to maintain a constant walking speed and spray height when using a hand-held or knapsack sprayer. Use a left-to-right sweeping motion with little overlap of the spray for single nozzle wands. Apply just enough solution to wet the leaf tissue.

Do not saturate leaf tissue or make dripping wet. This is the biggest mistake made when using hand-held or knapsack sprayers. Turfgrass injury can result.

To calibrate a hand-held or knapsack sprayer for applying pesticides, first fill your sprayer with a known amount of water. Spray an area of known size (*e.g.*, 20 ft x 50 ft = 1,000 ft2). Be sure you keep your spray pressure up by regularly pumping your sprayer (perhaps a single pump of the handle per step). After spraying the area, measure the amount of water remaining in the sprayer. Subtracting the remaining amount of water from the initial amount of water in the sprayer equals the spray volume per 1,000 ft2 (usually measured in gallons). Then calculate the amount of pesticide required per 1,000 ft2 (many labels give rates in ounces of product per 1,000 ft2) and add that amount of pesticide to the calibrated amount of water to cover 1,000 ft2. If spray area is less than 1,000 ft2, adjust spray mixture accordingly.

Self-propelled or pull-behind sprayers: Check your spray boom to make sure all strainers are clean and of the same type. Make sure all tips are the same type, angle and size. Most tips are now colour coded to make this task a little easier. Make sure the tank filter is clean and free of any blockages.

Fill your spray tank with an adequate amount (half tank) of water to check flow rate and set pressure. First, turn on your sprayer and check for leaks. Now is the time to make sure your spray system is working properly and make repairs as needed. If everything is operating properly, you are now ready to measure the flow rate.

Calibrating Spreaders

When calibrating spreaders, it is recommended to calibrate the spreader for half the rate desired. Applications of fertilizers or pesticides will be more evenly distributed if applied in two directions (x or + pattern). This also helps in preventing the striping patterns evident in poor application techniques where proper overlap was ignored. Therefore, calibrating for half the rate allows for applications in two directions giving a full rate of fertilizer or pesticide.

Drop Spreaders (Worksheet 2): Calibrating drop spreaders can be completed two different ways. Some drop spreaders have collection pans for catching dropped material, usually attaching to the bottom of your spreader. If your drop spreader does not have a collection pan, it is easy to make one and attach it with bungee cords. The collection pan should be oversized to collect all material dropped, but not very deep. This allows the pan to hang between the bottom of the spreader and the ground.

Method-1: First, determine from the product label the amount of granular pesticide, granular fertilizer, or a combination product you require per 1,000 ft2. Measure the width of your spreader in feet. Divide 100 by the width of your spreader and measure this distance with markers. (Example: 100 divided by 1.5 feet [18" wide] = 66.67 feet). The area measured represents 100 square feet, or one-tenth of the 1,000 square feet for one swath. Fill your spreader with the material you wish to calibrate. Set your spreader to the manufacturer's recommended setting or a midsize opening. Push the spreader at a normal pace in an area similar to that being treated, turning the spreader on and off for the distance marked. Weigh the material collected, which should equal one-tenth of the material required for 1,000 ft2. Adjust spreader and make several runs until the desired amount is collected.

Method-2: The second method requires the applicator to push the spreader over a collection pan or box (1 foot by the width of the spreader) on a smooth surface. A spreader that drops an 18-inch swath will require a collection box 1 foot x 1.5 foot (1.5 ft2). The applicator needs to know the amount of material required for one square foot. Set your spreader to the manufacturer's recommended setting or a midsize opening. You can make several passes over the collection box, in the same direction, to collect enough material to weigh. Check the distribution of material in the collection box to make sure an even amount of material is dropping across the entire width of the spreader. If material is falling unevenly, make sure there are no blockages in the bottom of your spreader. Divide the weight of the collected material by the number of passes, then divide by the number of square feet in the collection box. The result should be the amount of material required for one square foot. Adjust spreader and make several runs until the desired amount is collected. Be sure to clean-up all materials that fall outside of the collection box.

Rotary-type spreaders (Worksheet 3): Some cyclone spreaders have collection pans that can be attached around the spreader to collect all material thrown. Follow the first procedure for drop spreaders, if you have a collection unit. However, this does not allow you to determine the effective swath and overlap required. The following procedure calibrates the spreader for the material being used and determines how much overlap is needed.

Calibrating a cyclone spreader without a collection unit will require you to construct a series of collection pans or boxes 1' x 1' x 3" high (usually 11 to 13 boxes). This method also requires that the calibration be done on a smooth surface, because the actual material (granular pesticide or fertilizer) must be used. This allows you to cleanup any material that is thrown outside of the collection boxes.

Determine how much material the spreader deposits in each of 11 1-foot-square boxes. Calculate the distribution of material and adjust overlap of successive passes to achieve even coverage. The distribution shown here requires an overlap of 2 feet.

Determine how much material is required per 1,000 ft2 from the product label. Divide this amount by 1,000 to calculate the amount of material required for one square foot.

Set the spreader to the manufacturer's recommended setting or a midsize opening. Make several passes over the collection boxes, in the same direction, to collect enough material in each box for weighing. Divide the weight of the material collected in each box by the number of passes to equal the average amount of material thrown per square foot in each pass. Adjust the spreader and make several runs until the desired amount is collected.

Record the amounts of material from each box to see if you have an even distribution across the swath. Typically, there should be less material in the outer boxes. This determines how much overlap is required. For example, if the boxes contain the percentages of required material, then overlap should be 2 feet.

If your spreader throws more material to one side than the other, then the pattern plate at the bottom of your hopper should be adjusted. You may need to adjust your overlap if your spreader does not have a pattern adjustment and make all applications in the same direction. As always, be sure to cleanup all materials that fall outside of the collection boxes.

Again, an additional recommendation we make for spreader calibration, is to calibrate materials for half the amount required per one square foot. Then apply material in two directions (x or + pattern) for better distribution and elimination of striping.

Any applicator who becomes familiar with these calibrating procedures can make proper pesticide and fertilizer applications to achieve the best in pest control and plant growth.

9

Lawn Pesticide Facts

Pesticide Usage

- 78 million households in the U.S. use home and garden pesticides.
- Herbicides account for the highest usage of pesticides in the home and garden sector with over 90 million pounds applied on lawns and gardens per year.
- Suburban lawns and gardens receive more pesticide applications per acre (3.2-9.8 lbs) than agriculture (2.7 lbs per acre on average).
- Pesticide sales by the chemical industry average $9.3 billion. Annual sales of the landscape industry are over $35 billion.
- Included in the most commonly used pesticides per pounds per year are: 2,4-D (8-11 million), Glyphosate (5-8 million), MCPP (Mecoprop) (4-6 million), Pendimethalin (3-6 million), Dicamba (2-4 million).
- A 2004 national survey reveals that 5 million homeowners use only organic lawn practices and products and 35 million people use both toxic and non-toxic materials.

Health and Exposure Risks

- Of 30 commonly used lawn pesticides 19 have studies pointing toward carcinogens, 13 are linked with birth defects, 21 with reproductive effects, 15 with neurotoxicity, 26 with liver or kidney damage, 27 are sensitizers and/ or irritants, and 11 have the potential to disrupt the endocrine (hormonal) system.

- Pregnant women, infants and children, the aged and the chronically ill are at greatest risk from pesticide exposure and chemically induced immune-suppression, which can increase susceptibility to cancer.
- Scientific studies find pesticide residues such as the weedkiller 2,4-D and the insecticide carbaryl inside homes, due to drift and track-in, where they contaminate air, dust, surfaces and carpets and expose children at levels ten times higher than preapplication levels.

Children and Pesticides

- Children take in more pesticides relative to body weight than adults and have developing organ systems that make them more vulnerable and less able to detoxify toxins.
- The National Academy of Sciences estimates 50% of lifetime pesticide exposure occurs during the first 5 years of life.
- A study published in the Journal of the National Cancer Institute finds home and garden pesticide use can increase the risk of childhood leukemia by almost seven times.
- Studies show low levels of exposure to actual lawn pesticide products are linked to increased rates of miscarriage, and suppression of the nervous, endocrine, and immune systems.
- Exposure to home and garden pesticides can increase a child's likelihood of developing asthma.
- Studies link pesticides with hyperactivity, developmental delays, behavioural disorders, and motor dysfunction.
- Children ages children ages 6-11 have higher levels of lawn chemicals in their blood than all other age categories. Biomonitoring studies find that pesticides pass from mother to child through umbilical cord blood and breast milk.

Wildlife, Pets and Pesticides

- Studies find that dogs exposed to herbicide-treated lawns and gardens can double their chance of developing canine lymphoma and may increase the risk of bladder cancer in certain breeds by four to seven times.

- Of 30 commonly used lawn pesticides: 16 are toxic to birds, 24 are toxic to fish and aquatic organisms, and 11 are deadly to bees.
- Pesticides can be toxic to wildlife and cause food source contamination, behavioural abnormalities that interfere with survival, and death.
- Lawn and garden pesticides are deadly to non-target species and can harm beneficial insects and soil microorganisms essential to a naturally healthy lawn.

Pesticides in the Water

- Of 30 commonly used lawn pesticides, 17 are detected in ground-water, and 23 have the potential to leach.
- Runoff has resulted in a widespread presence of pesticides in streams and ground-water. 2,4-D, found in weed and feed and other lawn products, is the herbicide most frequently detected in streams and shallow ground water from urban lawns.
- Of the 50 chemicals on EPA's list of unregulated drinking water contaminants, several are lawn chemicals including herbicides diazinon, diuron, naphthalene, and various triazines such as atrazine.
- Runoff from synthetic chemical fertilizers pollutes streams and lakes and causes algae blooms, depleted oxygen and damage to aquatic life.

The Registration System and Pesticide Regulation

- The health data assessed by EPA for the registration of pesticides comes from the manufacturer of the pesticide. EPA is not obligated under the Federal Insecticide Fungicide and Rodenticide Act (FIFRA) to review peer-reviewed scientific literature.
- The U.S. GAO has told Congress on several occasions that the public is misled on pesticide safety by statements characterizing pesticides as "safe" or "harmless." EPA states that no pesticide is 100 percent safe.

- Pesticide testing protocol was developed before science fully understood the human immune and hormonal system. EPA still does not evaluate data for several neurological effects or disruption of the endocrine (hormonal) system.
- EPA does not evaluate the health and environmental effects of actual pesticide formulations sold on the shelf. Data submitted to the EPA also does not account for low-dose effects, synergistic effects with inerts or combined exposure to more than one pesticide at a time.
- Most states have preemption laws that prohibit localities from passing local pesticide-related ordinances that are stricter than the state policy.

Inert" Ingredients

- Pesticide products are made of an active ingredient and several inert, or other, ingredients. Inert ingredients are neither chemically, biologically nor toxicologically inert. Inerts are not disclosed to the public due to their status as "trade secrets".
- Active ingredients usually comprise only 5% of the actual product; the other ingredients make up the majority of a given pesticide product or formulation.
- Inert ingredients can be more toxic to humans than the active ingredient. Ethylene chloride, a nerve poison, is an example of an inert ingredient linked with damage to the heart, eyes, liver, and adrenal glands.
- 800 out of 1200 inerts are classified as "of unknown toxicity," 57 as highly toxic due to known carcinogenicity, adverse reproductive effects, birth defects, neurotoxicity and/or other chronic effects, and 64 as potentially toxic.
- 394 chemicals used as inert ingredients are listed as active ingredients in other pesticide products, and more than 200 inerts are considered hazardous pollutants and/or hazardous waste under federal environmental statutes.

Lawn Maintenance

Maintaining a healthy lawn involves using good maintenance practices throughout the growing season. These include proper mowing, fertility, irrigation, thatch control, integrated pest management, weed management and insect management. These techniques provide a dense, healthy, high quality lawn.

Mowing

The importance of good mowing practices is often overlooked but has a major influence on turf density, uniformity and aesthetic quality of a home lawn. It is the most repetitious and time-consuming maintenance practice and it is often done incorrectly.

Frequency and Height: Turf can be mowed frequently, provided no more than one-third of the grass blade is removed in a single mowing. Mow as high as possible. Lower mowing produces a shallow root system. Shallow grass roots cannot take up water and nutrients and make the lawn susceptible to drought stress.

Low mowing encourages broadleaf weed invasion and invasion from grassy weeds such as creeping bentgrass and annual bluegrass. It is best to mow a lawn when the leaves are dry. Dry grass cuts cleanly and clippings distribute more evenly.

Clippings: Leave clippings on. If they are excessively heavy, rake them up and remove them. Clippings contain nutrients and water, breakdown rapidly and do not contribute significantly to thatch. You can cut down your fertilizer (especially nitrogen) by 20%-35% by leaving the clippings on.

Mowers: Mowers need the capacity and power to handle the area being mowed. Consider weight, manoeuverability, ease in starting, ease of adjustment of height of cut and safety features. Keep mower blades sharp for a good quality cut. Select mulching type mowers that recycle grass clippings.

Fertility

Understanding and implementing a well-balanced fertilizer programme is one of the most important factors in maintaining an attractive healthy lawn. The three main nutrients required by lawns are:

- Nitrogen (N)
- Phosphorus (P)
- Potassium (K)

Nitrogen promotes dark green colour, leaf and blade development, and density of the turf. Phosphorus is important for good root and rhizome development and promoting plant maturity. Potassium contributes to the general vigour of the plant, promotes wear and drought tolerance and hardiness against winter injury.

Amount of Nutrients: The amount of nutrients required by a home lawn is best determined by soil testing. A soil test will provide levels of phosphorus, potassium, pH and lime requirements. There is no soil test for nitrogen. Generally, 1.5-2 kg/100 m^2 of actual nitrogen can be applied throughout the season, split into 2-4 applications. In the absence of a soil test, a 4-1-2 ratio (N-P-K) such as 20-5-10 is recommended. The three numbers on the fertilizer bag represent the amount of N, P and K in that order.

For example the 20-5-10 fertilizer ratio listed above contains 20% N, 5% P and 10% K. Nitrogen has to be applied every year, while phosphorus and potassium are relatively stable in the soil. If the lawn is on sandy soil, higher potash or more frequent applications may be required because it may leach. On newly established lawns, higher levels of phosphorus and potash may be required.

Fertilizers: Common turf nitrogen fertilizers and their properties are listed.

Application: An even application of lawn fertilizers is very important in achieving a uniform green lawn. If using a drop type spreader, operate it the long way of the lawn. First apply header strips at each end of the lawn to provide room for turning. Overlap one wheel's width when spreading the fertilizer and shut off the spreader when reaching the header strips. With a centrifugal type spreader, make two split (half rate each) applications at right angle to each other. Always make sure the spreader is properly adjusted, otherwise you could end up with striping or uneven colour.

Common Turf Nitrogen Fertilizers and Their Properties					
Type	**Examples**	**Response**	**Problems**	**Water Solubility**	**Potential for burn**
Inorganic	• ammonium nitrate • ammonium sulfate	• immediately available • quick green up	• readily leaches • causes lush growth	**High**	**High to very high**
Organic	• activated sewage sludge • animal by-products • plant by-products	• slow release • low leaching	• more expensive than inorganic forms	**Low**	**Low**
Synthetic	• IBDU • urea formeldehyde • sulfur coated urea • polymer coated urea	• slow release • low leaching	• release rate is dependent on temperature or moisture levels depending on the source	**Very low**	**Low to very low**

Irrigation

When normal rainfall does not provide enough moisture during the growing season grass goes dormant and turns brown. To ensure a high quality lawn the lawn must be watered. Signs that a lawn needs water include:

- footprints remain while walking across the lawn
- a slight change in colour to dark blue-green and grass blades folding inwards.

Frequency and Timing: Water in the early morning when there is little or no wind. This allows more even water distribution. Water before mid-day when the evaporation rate is the highest. Watering can be done in the evening, but this may encourage disease developing. Most disease-causing fungi require several hours of leaf wetness for infections and disease to occur.

Amount of Water: Too much water can cause thatch, fertilizer leaching, increased disease or grass weed problems such as creeping bentgrass, annual bluegrass or rough bluegrass. Too little water can cause shallow rooting of the turf, which makes the lawn susceptible to disease, drought stress or winter injury. Infrequent, thorough watering is best. When the lawn wilts, wet the entire area to a depth of 10-20 cm. The amount of water required to achieve this depends on soil characteristics. To measure how much water has been applied, place a straight-sided can or jar in the area being watered, and then check the water level in the can or jar. Approximately 2.5-4 cm of water in the can corresponds to an adequate irrigation of the lawn. Areas of the lawn needing more water include slopes, areas near buildings, curbs, sidewalks and light soils. Low lying areas, shaded areas and heavy soils may not need as frequent irrigation.

Dormant Lawns: During extended dry periods, a lawn may turn brown and go dormant. A lawn can survive from 4-6 weeks in a dormant state during summer dry periods. Once the rains return the lawn will green up in 7-10 days. If the lawn is dormant:

- Check regularly for insect pests
- Keep traffic off
- Stop mowing
- Do not fertilize

Irrigation Equipment: Hose watering is suitable for small areas only. A sprinkler attachment provides adequate coverage for an average size lawn despite the inconvenience in moving the sprinkler and a lot of water can be wasted. An underground irrigation system is the most expensive, but also the most efficient method, and may be considered for very large lawns or industrial properties.

Thatch

Thatch is a layer of organic matter made up of decaying grass leaves, stems and roots that build up in between the lawn and soil surface. It is a common problem on home lawns, particularly for lawns established for several years.

Identification: A thatchy lawn feels very spongy when walked on. Cut a triangular patch of lawn with a sharp knife and lift it back to measure the thickness of the thatch layer. More than 2.5 cm of thatch is too much.

Why is Thatch a Problem?: Thatch harbours insects and diseases. Thatch can restrict grass roots from growing into the soil root zone, resulting in a shallow rooted lawn. Thatch interferes with water infiltration.

Minimizing Thatch: Cultural practices that minimize thatch development are:

- frequent mowing
- proper watering
- proper ferilization

Excess thatch can be removed by vertical mowing or core aerating. Core aerate using a hollow steel tine core aerator which removes cores of soil. This physically breaks up the thatch, brings up beneficial soil micro-organisms that help break down the thatch and alleviates compaction.

Timing: Dethatch or aerate in spring and fall during periods of good growth, allowing for quick lawn recovery.

Overseeding: Overseeding is a method of thickening up a lawn that has become thin due to insects, disease or other types of damage. To insure success, add compost, peat or topsoil before overseeding. Overseed at double the seeding rate for establishing a new lawn. The best time to overseed a lawn is in the fall (mid-August to mid-September). Keep the overseeded area moist by watering several times a day. One week after seeding, reduce watering to twice a day until seedlings are established.

Intcgrated Pest Management

Try to manage pest problems in lawns by following the principles of integrated pest management. With integrated pest management prevention of pest problems is the best approach. Use good maintenance practices to promote a healthy lawn. It involves:

- Proper lawn planning and maintenance to prevent pests.
- Checking frequently for lawn pests.
- Proper identification of lawn pests.
- Treating with a pesticide only when a pest is present in large enough numbers to cause damage to the lawn.

The goal is to reduce pesticide use but still maintain healthy lawns.

Weed Management

A thick vigorous lawn is the best prevention against weed invasion. A dense stand of turf can compete successfully with weed seedlings for light and nutrients.

Weed Invasion: Problem weeds include both broad-leaved and grass weeds. They may occur when there are thin or damaged areas or heavily trafficked areas. Before attempting any weed control, weeds should be properly identified. Homeowners can consult nurseries, lawn care professionals, Publication 529, Weed Control in Lawns and Gardens or Publication 505, Ontario Weeds from the Ontario Ministry of Agriculture and Food or the Pest Diagnostic Clinic at the University of Guelph.

Control: Control problem weeds by:

- hand pulling
- hand raking
- mowing to prevent seed formation
- spot application of selective herbicides
- using a combination of the methods mentioned above.

Control a few weeds in a lawn by hand weeding or spot treating with a broad-leaved herbicide. If the weed infestation is too big to be handled with these methods, a broadcast herbicide may be required. Avoid using combination products containing herbicides and fertilizers unless the whole lawn is infested with weeds.

Most broadleaf herbicides available to home owners contain a combination of 2,4-D, mecoprop and dicamba. Weeds are easiest to control with herbicides when they are growing rapidly.

Herbicide Use: Spot treat whenever possible. Early spring or fall treatments are usually best. Younger weeds are more susceptible than mature weeds. Use the following precautions when applying herbicides:

- Do not apply during hot days because they may injure the lawn.

- Do not apply during the rain. Broad-leaved herbicides require 24 hours of dry weather to ensure good uptake by the plant.
- Delay mowing or irrigation for several days after application.
- Do not spray under windy conditions.
- Do not use 2,4-D or dicamba products on newly seeded lawns.

Preemergence herbicides are used to control weeds like crabgrass. To be effective, apply the herbicide before germination of weed seedlings. Only apply preemergence herbicides to areas with a history of a crabgrass problem. Read manufacturer's label for specific instructions on application and the required waiting period before re-seeding.

Insect Management

Insects that infest home lawns are generally difficult to notice and their presence goes undetected until significant damage has been done. Insect damage can often be mistaken for drought damage.

If the lawn remains brown or shows signs of thinning out despite watering, then closer examination for insects should be done. Regular inspection of the lawn including leaves, stems, roots, thatch and soil will help to determine if the problem is insect-related and catch the problem before extensive damage occurs. The most common lawn insect pests are:

- Hairy chinch bug
- Grubs
- Sod webworms
- European cranefly
- Bluegrass billbug
- Turfgrass Scale

Hairy Chinch Bug: Hairy chinch bug damage starts to become noticeable from mid-July to mid-August. It causes damage by sucking plant juice from the grass stems with its piercing and sucking mouth parts.

The first signs of damage are small fist sized sunken areas in the lawn. As the insects continue to feed these areas can grow very rapidly into large dead areas. Damaged areas are often taken over by weeds. Adults are very small (4 mm long) and their wings form an X on their backs. One method of detecting chinch bugs is to cut the ends of a metal can to make a cylinder and force it into the ground. Fill the cylinder with water, chinch bugs will float to the surface if they are present in the lawn. If chinch bugs are found in a lawn apply one of the insect pesticides available from garden centres, etc. that are labelled to control hairy chinch bugs.

Grubs: Several species of grubs can cause damage by feeding on grass roots in home lawns. The most common grubs are:

- June beetle
- European chafer
- Japanese beetle

If animals such as skunks begin digging up the lawn or if small irregular patches of grass turn brown, it may indicate grubs present in the lawn. Check for grubs in late summer or early fall before damage is noticed. Lift a section of sod and soil with a spade, 10 cm square and look for white, C-shaped grubs in the soil. Keeping the lawn well watered helps minimize grub damage. Insecticides that control grubs are available at local garden centres. Timing for grub insecticide treatments is late summer or early fall before extensive damage occurs. The insecticide must be watered in with 2.5 cm of water to be effective.

Sod Webworms: Sod webworms are caterpillars that feed in the thatch on home lawns. The adults are small tan moths that fly around lawns at night. The caterpillars grow to 2 cm and are tan in colour. They have dark spots on their backs. They feed in the thatch causing damage that is similar to grub damage, where areas of turf can be lifted like a carpet. They also leave behind soft green fecal pellets in the areas they are feeding. The damage occurs in September. Use products containing carbaryl to control sod webworms.

Bluegrass Billbug: Bluegrass billbugs are in the weevil family. The adults are black and 5 mm long. Larvae are small white and legless with a brown head. Damage begins with small groups of plants turning yellow and dying. This occurs in mid-July to mid-August. Damage is usually spotty and rarely affects a whole lawn. Billbugs leave behind a sawdust-like excrement in areas they have been feeding. This is helpful in diagnosing the damage.

European Cranefly: The European cranefly is a relatively new pest to lawns in Ontario. The adult cranefly resembles a large mosquito. The larvae are known as leatherjackets. They are light greyish-brown with black specks. There is no visible head region. They range in size from 0.5-3.0 cm in length. Leatherjackets feed primarily on grass shoots during the evening and on grass roots during the day. Damage begins to show between early to mid-May and peaks by mid-June. Heavy infestations of leatherjackets can chew the grass down to the bare soil.

Turfgrass Scale: Turfgrass scale is a typical scale insect that resembles an egg cut in half lengthwise. It is brown with a yellow stripe in the middle. The immature stage is called a crawler and it is the size of the head of a pin. Typical damage is small patches of dead grass that do not green up in the spring. They are found mainly on sodded lawns and in general, do not cause much damage. During early July crawlers can be found on shoes when you walk through the lawn. There is no insecticide registered for use on turfgrass scale in Ontario.

Endophytes: One way to combat the damage caused by leaf feeding insects is to plant grasses that contain endophytes. Endophytes are fungi that grow inside the grass plant and make it taste bad. Lawn insects repelled by endophytes are hairy chinch bugs, bluegrass billbugs and sod webworms. A summary of the major turf insect pests, how to identify them, the damage they cause and how to treat them can be found. Lawn species that may contain endophytes are perennial ryegrass, fine fescues and tall fescue. Consult your local garden centre or seed supplier about the endophytic grasses they sell.

Other Pests

Moss: Mosses are primitive forms of green plants that form tangled clumps that grow in amongst a lawn. Moss plants have a tremendous capacity to spread. They produce large numbers of spores, each of which can give rise to a new moss plant. Mosses are poor competitors with a healthy lawn and are only a problem when the lawn is weakened by poor growing conditions. The following conditions that weaken a lawn and promote moss invasion:

- poor aeration
- poor drainage
- low fertility
- high acidity
- heavy shade

Cultural methods can be effective in controlling moss. They are:

- Adequate fertility, especially nitrogen (2.0 kg per 100 m^2 of actual N).
- Control thatch by vigorous raking, aeration or vertical mowing.
- Reduce shade by pruning trees.
- Maintain a soil pH of between 6 to 7 - apply lime if soil is acidic.
- Improve surface drainage.
- Reduce compaction and increase aeration by spiking slicing or aerifying with a mechanical aeifier.
- Do not over-water.
- Do not remove more that 1/3 of the top growth at one time when mowing and mow at a height appropriate for the grass species.

Disease Management

If integrated pest management practices are followed, disease problems in home lawns are minimal. Excessive fertility can cause succulent growth that is more susceptible to diseases. When establishing or renovating a lawn area, select varieties that are resistant to various diseases. Improper irrigation also contributes to lawn diseases. Make sure to water in the early morning to minimize the length of time that the lawn stays wet. A summary of common turfgrass diseases on home lawns.

Outlines a sample integrated pest management programme for home lawns in Ontario. It can be used as a guideline for fertilizing, overseeding and aerating as well as for pest monitoring and treatment.

Pesticide Use

The two types of pesticides commonly used on lawns are herbicides for weed control and insecticides for insect control. Before applying a pesticide follow the steps of integrated pest management. Make sure that the pest is properly identified. When purchasing a pesticide make sure the label includes the pest you are trying to kill and that the pesticide lists lawns. Read the label carefully and follow the manufacturer's directions. Spot treat areas where possible and time applications correctly to avoid repeat application.

Landscape Introduction

Lawns, ornamental shrubs and trees, often poorly adapted to difficult urban conditions, have for many years dominated urban landscape design. To survive, these landscapes require high levels of maintenance, irrigation, fertilizers, herbicides and pesticides. These not only increase the cost of ownership, but pollute stormwater runoff and provide little or no habitat for local wildlife. Ecologically based landscape design can change this pattern.

It considers a region's water flows, climate, open spaces, native vegetation, wildlife and existing urban development. Understanding the ecology of the region helps when working with the needs of a specific site design, including its solar access, wind patterns, topography, rainfall and drainage patterns, the preservation of existing vegetation and unique features of the site.

In recent years the City of Santa Monica has promoted a more environmentally responsive approach to landscape design, particularly in water conservation. Through the City requirement to use drought-resistant plants, restrict the amount of lawn and the use of efficient irrigation systems, water consumption has dropped significantly.

The Required Practices focus on landscape design that conserves water and improves water quality. Outdoor water use accounts for approximately 35% of overall water consumption in Greater Los Angeles, of which half is used to irrigate gardens and landscaping. Planting drought-tolerant vegetation, minimizing lawn areas and using drip irrigation can reduce water use in the garden by 50% to 70% and overall water consumption by 20% to 25%. By managing stormwater runoff through on-site surface water retention and reducing the area of impermeable paving, fewer pollutants enter Santa Monica Bay.

The Recommended Practices address a broader range of environmental benefits gained through landscape design. These include improving the microclimate around buildings to reduce cooling energy consumption and improve occupant comfort, reducing glare, enhancing the appeal of exterior spaces, providing habitat for urban wildlife and creating opportunities for local food production.

Controlling Solar Gain and Improving Ventilation: Controlling solar gain in a building through planting on south, southwest and southeast sides of buildings is a very effective way to reduce cooling loads. In Santa Monica, where southwest-oriented building facades have the greatest solar heat gain, planting shade trees regulates heat gain and reduces air-conditioning needs. Trees and other vegetation can also shade paved areas and provide a canopy for pedestrians.

In addition to controlling solar gain, landscaping and landscape structures can enhance a building's natural ventilation and cooling by influencing wind flow. This requires the landscape design to respond to prevailing wind patterns and local wind flow variations, and strategically locating trees, hedges and shrubs to direct winds toward ventilation inlets and create low-pressure areas at outlets.

Dissipating turbulent, higher velocity winds in circulation areas around buildings can also be achieved by placing a multi-layered planting of large trees and shrubs with dense foliage at the base and corners of buildings.

Air Quality and Wildlife Habitat : Ecologically based landscape design can improve local air quality by absorbing carbon dioxide, producing oxygen and filtering particulates. Landscaped parks and streets provide a "green lung" for the community. Building roofs and balconies can be deliberately designed to accommodate vegetation. Conventional roofs degrade air quality by replacing a site's original vegetative cover with an impervious surface that absorbs solar energy and contributes to the urban heat island effect. Planted roofs provide a living surface that cools and insulates the building, improves air quality and protects roof membranes from ultraviolet breakdown.

Landscape design can create important habitat for resident and migratory wildlife by providing food and shelter. Through mulching and composting green matter, ecologically-based landscaping can maintain and improve soil fertility, retain soil moisture, stimulate soil-based microbial activity and maintain the cycle of nutrients in the soil. In addition, the use of native and other ecologically appropriate plants reduces water consumption, maintenance requirements, and the use of pesticides and herbicides, and increases the area's suitability for wildlife.

Health Benefits of Ecological Landscaping: Finally, environmentally based landscape design practices can have direct financial and human health benefits. Capital and annual utility costs and regular maintenance requirements can be reduced, resulting in lower overall operating costs. Incorporating areas to grow food within developments encourages local food production and reduces monthly food expenses.

It can provide occupants with a secure source of organically grown, fresh produce. Opportunities to garden provide personal health benefits that can improve the quality of life, particularly for older people with leisure time to enjoy gardening.

Bibliography

Adams H.: *Hot Pepper Improvement*, St. Michael, CARDI, 1997.

Agarwal P. K.: *Improvement of Citrus*, New Delhi, Malhotra Publishing House, 1993.

Alexander M. P. and Ganeshan S.: *Pollen Storage*, New Delhi, Malhotra Publishing House, 1993.

Andrews J.: *The Domesticated Capsicums*, University of Texas Press. 1995.

Andrews L.: *Citrus Production - Orange*, St. Augustine, Trinidad and Tobago, 1990

————: *Production of Minor Crops Sapodilla, Tamarind, Guava and Genip*, Port of Spain, Trinidad and Tobago, 1996.

Arguello D. S.: *Production, Post-harvest Management and Exportation of Tropical Fruits in Costa Rica*, Wageningen, Technical Centre for Agricultural and Rural Cooperation, 1992.

Ashworth S.: *Seed to Seed*, Decorah, Seed Savers Publications, 1991.

Barbeau G.: *Tropical Fruits in Nicaragua*, Managua, Nicaragua Ministerio de Desarrollo Agropecuario, Agraria, 1990.

Batlle I. and Tous J.: *Carob Tree (Ceratonia siliqua L.)*. Rome, International Plant Genetic Resources Institute, 1997.

Berger, P.L.: *Pyramids of Sacrifice: Political Ethics and Social Change*, New York, Basic Books, 1974.

————: *Pyramids of Sacrifice: Political Ethics and Social Change*, New York, Basic Books, 1974.

Brooks, D.: *Water: Local-Level Management*, Ottawa, International Development Research Centre, 2002.

Bunnik J. S. C.: *Fresh Fruits and Vegetable: a Survey on the Netherlands and other Major Markets in the European*

Community, Netherlands, Centre for the Promotion of Imports from Developing Countries, 1990.

Burton W.G.: *The Potato*, Holland, H. Veenman & Zonen N.V., 1966.

Cao Van P.: *An Integrated Approach for the Production and Processing of Minor Fruits in Martinique (FWI)*, Port of Spain, Trinidad and Tobago, 1996.

Cardona M. J. G., Carvajal S. L. B., Salinas D. G. C. and Isaza R. G. B. : *Proceedings of the International Seminar on Plantain Production, Quindio, Colombia,* Quindio, Corporation Colombiana de Investigation Agropecuaria, 1998.

Chadha K. L. and Pareek O. P.: *Advances in Horticulture: Fruit Crops,* New Delhi, Malhotra Publishing House, 1993.

Chambers, R.: *Rural Development: Putting the Last First*, London, Longman, 1983.

Collymore L.: *Fruit Production in Barbados*, Port of Spain, Trinidad and Tobago, 1996.

Coste R.: *Coffee: the Plant and the Product*, London, MacMillan, 1992.

Crucefix D.: *Avocado Variety Selection for Export Development,* Roseau, CARDI, 1996.

Cull Brian & Lindsay Pax: *Fruit Growing in Warm Climates for Commercial Growers & Home Gardeners*, Australia, Reed Books, 1995.

Currah L. and Proctor F. J.: *Onions in Tropical Regions*, Kent, Natural Resources Institute, 1990.

Daniells J.: *Illustrated Guide to the Identification of Banana varieties in the South Pacific*, Canberra, ACIAR, 1995.

Degras L.: *Yam: a Tropical Root Crop*, Wageningen, CTA/ MacMillan, 1993.

Dhatt A. S. and Singh Z.: *Propagation and Rootstocks of Citrus*, New Delhi, Malhotra Publishers, 1993.

Diederichsen A.: *Coriander (Coriandrum sativum L.)*. Rome, International Plant Genetic Resources Institute, 1996.

Doijode S. D.: *Seed Germination in Fruits*, New Delhi, Malhotra Publishers, 1993.

Dudley, E.: *The Critical Villager: Beyond Community Participation*, London, Routledge, 1993.

Featherly H. I.: *Taxonomic Terminology of the Higher Plants*, USA, Iowa State College Press, 1954.

Ferentinos L.: *Proceeding of the Sustainable Taro Culture for the Pacific Conference*, Honolulu, HITAHR, 1993.

Forde S: *Proceedings of CARDI / CTA Workshop on Marketability of Caribbean Minor Fruits*, Port of Spain, Trinidad and Tobago, 1996.

Georges, S.: *The Debt Boomerang: How Third World Debt Harms Us All*, Boulder, Westview Press, 1992.

Glowinski Louis: *The Complete Book of Fruit Growing in Australia*, Australia, Lothian Publishing Company Pty. Ltd., 1991.

Godden G.: *Growing Citrus Trees*, Australia, Lothian Publishing Company Pvt. Ltd., 1988.

Gowen S.: *Banana and Plantains*, London, Chapman Hall, 1996.

Green S.K. and Kim J. S.: *Sources of Resistance to Viruses of Pepper (Capsicum spp.): a Catalogue*, Taipei, AVRDC, 1994.

Gunjate R. T. and Tawde A. B.: *Propagation and Rootstocks of Mango*, New Delhi, Malhotra, 1993.

Hartley W.: *A Checklist of Economic Plants in Australia*, Melbourne, C.S.I.R.O., 1979.

Harwood, R. R.: *Marshalling Technology for Development: Proceedings of a Symposium*, Washington, DC, National Academy Press, 1995.

Herklots G. A. C.: *Vegetables in South East Asia*, London, George Allen & Unwin Ltd., 1972.

Hessayon D. G. Dr.: *The Vegetable Expert*, England, PBI, Publications, 1985.

Huaman Z.: *Descriptors for Sweet Potato*, Rome, International Board for Plant Genetic Resources, 1991.

Hurst Jacqui & Rutherford Lyn.: *A Gourmet's Book of Mushrooms & Truffles*, Sydney, Golden Press Pvt. Ltd., 1991.

Jacquat Christiane: *Plants from the Markets of Thailand*, Bangkok, Duang Kamol, 1990.

Jeffers P.: *Evaluation of Four Onion Varieties in Montserrat*, Plymouth, CARDI, 1992.

Kenridge K. C. and Hardy B.: *Biology and Agronomy of Forage Arachis*, Cali, International Centre for Tropical Agriculture, 1994.

Kroll R.: *Cut Flowers*, Wageningen, CTA, 1995.

Kunelius T.: *Annual Ryegrasses in Atlantic Canada*, Ottawa, Agriculture Canada, 1991.

Lofgren, H., Richards, A.: *Food Security, Poverty, and Economic Policy in the Middle East and North Africa*, Washington, DC, International Food Policy Research Institute, 2003.

Mabberley D. J.: *The Plant-Book : a Portable Dictionary of the Vascular Plants*, Cambridge, Cambridge University Press, 1997.

Madulid Domingo A.: *A Pictorial Cyclopedia of Philippine Ornamental Plants*, Philippines, Makati Metro Manila, 1995.

Malins A.: *Postharvest Handling of Pineapple and Mango*, Port of Spain, Trinidad and Tobago, 1992.

Mannetje, L. T. & Jones, R. M.: *Plant Resources of South-East Asia,* Wageningen, Pudoc Scientific Publishers, 1992.

Matsuoka H.: *Cultivation of Panicum Genetic Resources for Evaluation of Characteristics*, Tokyo, JICA, 1997.

Mc Donald F.: *Plant Tissue Culture Manual for Yam, Cassava, Sweet Potato, Dasheen (taro) and Tannia (cocoyam)*, Roseau, Dominica and Cave Hill, 1993.

Mellor, J. W.: *The New Economics of Growth*, Ithaca, Cornell University Press, 1976.

Miller William: *Dictionary of English Plant Names*, London, John Murray, 1884.

Mitra S.: *Postharvest Physiology and Storage of Tropical and Subtropical Fruits*, Oxon, CABI, 1997.

Morris M. L. : *Maize Seed Industries in Developing Countries*, Colorado, Lynne Rienner, 1998.

Morton Julia F.: *Fruits of Warm Climates*, Miami, Julia F. Morton Publisher, 1987.

Murali T. P. and Duncan E. J.: *In Vitro Propagation of Banana through Aseptic Manipulation of Male Inflorescences*, Port of Spain, NIHERST, 1990.

Nabhan G. P.: *Wild Phaseolus Ecogeography in the Sierra Madre Occidental, Mexico: Areographic Techniques for Targeting and Conserving Species Diversity*, Rome, International Board for Plant Genetic Resources, 1990.

Nijdam J. & De Jong A.: *Elsevier's Dictionary of Horticulture in Nine Languages*, Elsevier Scientific Publishing Co. Amsterdam; New York.

Oka H. I.: *Origin of Cultivated Rice*, Elsevier, Japan Scientific Societies Press, 1988.

Oldham P.: *Cost of Production of Major Tree Crops in Dominica*, Roseau, Ministry of Agriculture, 1991.

Ou Yang Jue Ya: *Comparative Study of Mandarin and Cantonese*, Beijing, China Social Sciences Publishing House, 1993.

Percival John: *The Wheat Plant - A Monograph*, London, Duckworth & Co., 1921.

Pilgrim R.: *Post Harvest Handling of Minor Exotics*, St. George's, Grenada, 1996.

Ragone D.: *Breadfruit: Artocarpus Altilis (Parkinson) Fosberg*, Rome, International Plant Genetic Resources Institute, 1997.

Rehm Sigmund: *Multilingual Dictionary of Agronomic Plants*, Boston, Kluwer Academic Publishers, 1994.

Roy S. K.: *Research on Multipurpose Tree Species in Asia: In Vitro Clonal Propagation of Artocarpus Heterophyllus*, Bangkok, Winrock International Institute for Agricultural Development, 1991.

Singh H. P. and Chadha K. L.: *Genetic Resources of Citrus*, New Delhi, Malhotra Publishing House, 1993.

Sinha G. C., Reddy Y. T. N. and Singh G.: *Propagation and Rootstocks in Guava*, New Delhi, Malhotra Publishers, 1993.

Sperling L. and Berkowitz P. *Partners in selection: bean breeders and women bean experts in Rwanda*. Washington, USA: CGIAR, 1994.

Sreekumar V., Indrasenan G and Mammen G.: *Studies of the Quantitative and Qualitative Attributes of Ginger Cultivars*, Calicut, Kasaragod, 1990.

Stover R. H. and Simmonds N. W.: *Bananas*, United Kingdom: Longman Scientific and Technical, 1991.

Thavarasook C.: *Proceedings of the Mungbean Meeting 90, Chiang Mai, Thailand,* Bangkok, Tropical Agricultural Research Centre, 1991.

Thomas E.: *Fruit Production in St. Kitts and Nevis*, Port of Spain, IICA, 1996.

Vargas E. M., Macaya G., Baudoin J. P. and Rocha O. J.: *Variation in the Content of Phaseolin in Wild Populations of Lima Beans (Phaseolus lunatus L.) in the Central Valley of Costa Rica*. Plant Genetic Resources Newsletter, 2000.

Webb M.: *Weed Problems and their Control in Rice, Papaya, Citrus and Sugar cane: a Report to the Belize Plant Protection Service*, United Kingdom, Natural Resources Institute, 1993.

Whealy K.: *The Garden Seed Inventory*, Decorah, Seed Saver Publications, 1988.

Whitwell A.: *Dominica Orchard Crop Management and Research Project: Pest and Disease Management,* United Kingdom, Natural Resources Institute, 1991.

Woolfe Jennifer A.: *The Potato in the Human Diet*, Cambridge, Cambridge University Press, 1989.

Yoshida T.: *Cultivation of Citrus Genetic Resources for Evaluation of Characteristics*, Tokyo, JICA, 1996.

Index

A

B

C

D

E

F

R

S

T

❑❑❑